The Economics of Estuary Restoration in South Africa

This book examines the economic costs and benefits of the ecological restoration of estuaries, utilizing case studies from South Africa.

Estuaries are important ecosystems from both an ecological and human perspective. Yet, in many parts of the world they are often degraded environments, facing threats from climate change, invasive species, fire and wastewater pollution. While the environmental benefits of restoring degraded environments are well discussed, this book specifically examines the economic benefits of doing so. It applies a cost-benefit analysis, which focuses on a range of key ecosystem services, including human health, fishing value, recreational value and property value. The book utlizes three detailed studies of the Swartkops estuary, the Great Brak estuary and the Knysna estuary in South Africa, but also draws out lessons that can be applied to coastal environments across the world. Overall, this book demonstrates that ecological restoration does pay and that the value of additional ecosystem services gained through restoration far exceeds the costs associated with this restoration process.

This book will be of great interest to students and scholars of environmental management and restoration, ecological economics, ecosystem services and environmental conservation.

Douglas J. Crookes is an associate professor of environmental governance in the School for Public Leadership at Stellenbosch University, South Africa. He has over 22 years of experience conducting problem-focused and applied research at the interface between the environment and economics. He is the author of *Mathematical Models and Environmental Change: Case Studies in Long Term Management* (Routledge, 2022).

Routledge Focus on Environment and Sustainability

Regional Political Ecologies and Environmental Conflicts in India
Edited by Sarmistha Pattanaik and Amrita Sen

Circular Economy and the Law
Bringing Justice into the Frame
Feja Lesniewska and Katrien Steenmans

Land Tenure Reform in Sub-Saharan Africa
Interventions in Benin, Ethiopia, Rwanda, and Zimbabwe
Steven Lawry, Rebecca McLain, Margaret Rugadya, Gina Alvarado, and Tasha Heidenrich

Agricultural Digitization and Zhongyong Philosophy
Creating a Sustainable Circular Economy
Yiyan Chen, Hooi Hooi Lean, and Ye Li

EU Trade-Related Measures against Illegal Fishing
Policy Diffusion and Effectiveness in Thailand and Australia
Edited by Alin Kadfak, Kate Barclay, and Andrew M. Song

Food Cultures and Geographical Indications in Norway
Atle Wehn Hegnes

Sustainability and the Philosophy of Science
Jeffry L. Ramsey

Food Cooperatives in Turkey
Building Alternative Food Networks
Özlem Öz and Zühre Aksoy

The Economics of Estuary Restoration in South Africa
Douglas J. Crookes

For more information about this series, please visit: www.routledge.com/Routledge-Focus-on-Environment-and-Sustainability/book-series/RFES

The Economics of Estuary Restoration in South Africa

Douglas J. Crookes

First published 2024
by Routledge
4 Park Square, Milton Park, Abingdon, Oxon OX14 4RN

and by Routledge
605 Third Avenue, New York, NY 10158

Routledge is an imprint of the Taylor & Francis Group, an informa business

British Library Cataloguing-in-Publication Data
A catalogue record for this book is available from the British Library

ISBN: 978-1-032-65165-1 (hbk)
ISBN: 978-1-032-65169-9 (pbk)
ISBN: 978-1-032-65167-5 (ebk)

DOI: 10.4324/9781032651675

Typeset in Times New Roman
by MPS Limited, Dehradun

Contents

Acknowledgements

Most of this book was written while the author was based at Nelson Mandela University's Institute for Coastal and Marine Research (CMR). I would like to thank One Ocean Hub for providing funding for me while I was at this institute. The author would like to thank the following people for assisting and providing inputs throughout the development of the book. First, I thank my editor at Taylor and Francis, Hannah Ferguson, for guidance and support throughout the process, and Katie Stokes, also from Taylor and Francis, for excellent editorial assistance. In terms of specific chapters, I would like to thank the following people for their comments and inputs: Janine Adams (NMU), Daniel Lemley (NMU), Susan Taljaard (CSIR), Paula Melariri (NMU) and Bernadette Snow (University of Strathclyde) (Chapter 3); James Blignaut (SU) and Myles Mander (Chapter 6). Editorial assistance from Leandri van der Elst is gratefully acknowledged. I thank three anonymous reviewers for comments, and thank my current institution (Stellenbosch University) for research support. Any errors or omissions that remain are solely those of the author.

1 Introduction

1.1 Background and definitions

The estuarine realm is one of the most endangered ecosystem types in South Africa (Van Niekerk et al. 2019). A recent National Biodiversity Assessment for estuaries (2018) found that "multiple interventions are required to avoid further decline in health. These include protection of freshwater inflow, restoration of water quality, reduction in fishing effort and avoidance of mining, infrastructure development and crops in the EFZ" (NBA 2019, p. 30).

Estuaries provide a number of important economic benefits to communities living adjacent to them. These livelihoods are likely to be impacted by estuary degradation. Imperatives such as ecological restoration and ocean science are global imperatives, as highlighted by the United Nations (UN) – which recently announced both the UN decade of ecological restoration and the UN decade of ocean sciences (2021–2031).

It is on the basis of this that an assessment of the economics of estuary restoration is conducted, considering cases studies from specific estuaries in South Africa. In this introduction, definitions of natural capital and ecosystem restoration are considered, along with the benefits of ecological restoration, the economics of estuaries and the threats to estuaries.

1.1.1 What is natural capital?

Natural capital is defined by Daly and Cobb (1989:72) as "the non-produced means of producing a flow of natural resources and services". According to Aronson et al. (2007), there are the following four components to natural capital:

- Renewable natural capital, that is, the ecosystem and living species contained therein

DOI: 10.4324/9781032651675-1

- Non-renewable natural capital, that is, assets in the subsoil, for example, coal, oil and gold
- Replenishable natural capital, for example, water resources, atmosphere and fertile soils
- Cultivated natural capital, for example, crops, plantations and orchards

It is evident from this that natural capital is the physical stock of natural assets. However, it affects human well-being through the flow of goods and services. The restoration of natural capital is "any activity that integrates investment in and replenishment of natural capital stocks to improve the flows of ecosystem goods and services, while enhancing all aspects of human well-being" (MA 2005:5). It is generally understood that the Millennium Ecosystem Assessment (2005) constituents of well-being are the following:

- Security (including personal safety, secure resource access and freedom from natural disasters)
- Basic material for a good life (including livelihood sufficiency, nutrition and access to goods and services)
- Health (including physical strength and access to clean water, air and sanitation)
- Good social relations (including social cohesion, mutual respect and the ability to help others)
- Freedom of choice and action (undergirding all of the above)

The interaction between human well-being and ecosystems is evident by the fact that the natural environment provides many of the resources necessary for human survival, fulfilment and enjoyment. In addition, ecosystems are important contributors to the real economy (De Wit et al. 2012). This interdependence between natural and economic systems is usually highest in developing countries, even though the economic values associated with natural resources may be higher in developed countries due to higher disposable incomes. In addition, economic systems impact natural systems through the generation of waste and heat (Costanza 2001). Natural capital may provide a means of recycling or assimilating these wastes (Costanza and Cleveland 2008).

1.1.2 Benefits of ecological restoration

The economic arguments for ecological restoration are quite compelling. In a review of five case studies, Balmford et al. (2002) found that sustainable management of natural resources across a range of biomes resulted in gains in Net Present Values (NPVs) of between 14% and 75% compared with unsustainable practices. While there have been a number of success stories in restoring natural capital internationally (for a discussion of the famous New York City-Catskills project, see Elliman and Berry 2007), the results from individual studies on benefits and costs are mixed. On the benefit side, Tong et al. (2007), for example, concluded that there is a significant potential gain. They estimated an increase of 89.5% in ecosystem service value as a result of the restoration of the Sanyang wetland in China. On the other hand, costs of restoration are high, and opportunity costs may often be a significant barrier to conservation on private lands (Dorrough et al. 2008).

Results of a range of local and international studies indicate that, while the economic value of estuaries is high (Table 1.1), ecosystem values that include the costs of restoration are substantially lower (yet still positive). However, some high values are possible. In a study in Bushbuckridge, Blignaut and Moolman (2006) found a potential net gain in the direct consumptive use benefit of restoration of US$391/hectare, or US$ 72 million across all the land under communal management, but this value is not annualized and may therefore not be comparable to the other values in the Table. Higgins et al. (1997) developed a model for mountain fynbos

Table 1.1 Economic values for different ecosystem services

	Type of value	*Where?*	*Value of benefit (US $/ha/yr)*	*Notes*
Estuary values				
Turpie et al. (2010)	Existence value	Knysna estuary	457	2005 values[1]
Costanza et al. (2006)	Aesthetics/ recreation	New Jersey (estuary), USA	748	2004 values[1]
De Groot et al. (2010)	Nursery (estuaries)	Various	37	2013 values[1,3]
De Groot et al. (2010)	Food (estuaries)	Various	129 (median)	2013 values[1,3]
Non-estuary				
Crookes (2023a)	Fisheries (marine)	Western Cape	7.67	2021 values, 80 years, 6% discount rate[2]
Crookes (2018)	Water saved (catchment)	Western Cape	11.44	Annualized, 2017 values[2]
Vundla et al (2017)	Water saved (catchment)	KwaZulu-Natal	90.23	Annualized, 2017 values[2]

Notes

1 Excludes restoration cost

2 Includes restoration cost

3 These values were taken from Blignaut et al. (2017)

dynamics with five sub-components (namely hydrological, fire, plant, management and economic valuation) and six value components (namely hiker visitation, ecotourist visitation, genetic storage, endemic species, wildflower harvest and water production). They estimated that the cost of restoration could range between 0.6% and 4.76% of total value, depending on the economic valuation scenario.

A preliminary review of cost-benefit studies suggests some promising results. During a study in the subtropical thicket in the Eastern Cape, Mills et al. (2007) found that the financial benefits are potentially positive, with an internal rate of return (IRR) of 9.2%. Holmes et al. (2004) also found a positive NPV in favour of both partial and full restoration of riparian habitat in North Carolina.

Several meta-analyses also found substantial economic benefits from ecological restoration. Aronson et al. (2010) reviewed restoration cases internationally, and Crookes and Blignaut (2019) reviewed 37 cases in South Africa, and more recently Peacock et al. (2023) also conducted a review of restoration case studies. All found that "restoration pays".

Table 1.2 summarizes some of the long-term economic benefits from environmental restoration. These include recreation benefits, fishery benefits, water quality improvements, erosion control and many others.

Table 1.2 Long-term economic benefits of environmental restoration

Benefit	*Source*
Aesthetics	
Increased property values	Bark et al. 2009; Isley et al. 2011; Kiel and Zabel 2001
Increased tourism	Isley et al. 2011; McCormick et al. 2010
Recreation	
Boating, swimming, water sports	Carson and Mitchell 1993
Park visitation	McCormick et al. 2010
Fish and game	
Fishery enhancement	Barbier 2007; Kroeger 2012; Kruse and Scholz 2006
Wildlife enhancement	Vickerman 2013
Ecosystem services	
Erosion control	Kroeger 2012
Stormwater management	Valderrama et al. 2013
Groundwater recharge	McCormick et al. 2010
Surface water availability	Milon and Scrogin 2006; Mueller et al. 2013
Water quality	Kroeger 2012; Milon and Scrogin 2006; Vickerman 2013
Flood control	Barbier 2007; Kroeger 2012; Milon and Scrogin 2006
Carbon sequestration	Vickerman 2013, Weinerman et al. 2012

Source: BenDor et al. (2015)

1.1.3 Economics of estuaries

Estuaries provide important economic benefits to communities living adjacent to them. In a study by Cooper et al. (2003), a number of economic benefits associated with eight estuaries in South Africa were assessed. Table 1.3 summarizes the results. Although they only report on three of the potential use benefits of estuaries (fishing values, recreational values, and property values), and for only eight of the 300 estuaries in South Africa, it is evident from this that the benefits of estuaries in South Africa is substantial (>R3 billion in 2002 prices, or US$170 million).

More recently, Turpie et al. (2017) conducted an assessment of the economic benefits of estuaries for the whole of the South African coastline. Benefits included fishing support, property premium, and provision of harvested resources. They estimated that the value of subsistence harvesting from estuarine habitats was R35.7 million per year; the nursery value of estuaries was estimated to be R803 million per annum. The property price premium for living on the coast in South Africa was estimated to be R79 billion per annum. Although some of this value may include coastal areas that are not estuaries, and therefore the property value component of estuaries may be overstated somewhat, the total value of estuaries by their estimation is nonetheless substantial (in the order of R80 billion, or US$4.5 billion per annum). At the same time, degradation of estuaries can reduce this value. These same authors estimate that almost half (42%) of the value of estuaries has been lost due to degradation of these systems.

Table 1.3 Lower bound estimates of the economic value associated with the sampled estuaries in South Africa (in 2002 Rands)

Values	*Fishing value*	*Recreational value*	*Property values*	*Total*
Berg	2,611,279	331,503,403	11,850,000	345,964,682
Breede	12,990,873	414,455,094	50,000,000	477,445,967
Umhlatuze/ Richard's Bay	12,550,466	6,870,264	Not applicable	19,420,730
Keiskamma	40,346,353	53,964,106	1,000,000	95,310,459
Orange	7,526,244	Not determined	Not determined	7,526,244
Swartkops	55,700,836	Not determined	Not determined	55,700,836
Knysna	46,624,387	Not determined	2,000,000,000	2,046,624,387
Kosi Bay	15,576,820	Not determined	Not determined	15,576,820

Source: Cooper et al. (2003).

This highlights the importance, from an economic perspective, of restoring these ecosystems.

1.1.4 Causes and consequences of degradation in estuaries

Although the economics studies mentioned in the previous section are important, and serve to highlight a number of important values associated with estuaries in South Africa, Van Niekerk et al. (2019) highlight a number of impacts on estuaries that are not included in the above estimates of economic value. These include impacts on estuaries due to invasive alien species, impacts on macrophyte species due to estuary degradation, and impacts on estuaries caused by nutrient enrichment (see also Adams et al. 2020 in this regard). Furthermore, fire can also affect estuarine systems, limiting the provision of ecosystem services (Barros et al. 2022). The presence of invasive alien plants can increase the intensity of fires (Brooks et al. 2004), thereby worsening effects on estuaries. This affects the restoration regime that is implemented for estuaries.

Although some studies have incorporated the costs of environmental degradation in estuary values (e.g. Nahman, 2006; Nahman and Rigby, 2008, and see Table 1.4 for an assessment of the Kongweni estuary in

Table 1.4 Summary of values, costs of degradation and benefits of restoration, for Kongweni estuary (Rand million, annual 2019 values)

	Kongweni estuary	*Comment*
Value	627	Fisheries, marine, biodiversity protection, streamflow regulation, water quality improvement, recreational values
Cost of degradation	127.6	Lower bound. Change in visitor spending as a result of a decline in water quality
(% of value)	20%	
Net value	499.4	
Restoration:		
A. Improved water quality (10%)	136.4	Change in visitor spending as a result of water quality improvement
B. Removal of invasive aquatic plant species (value of seagrass recovery)	n/a	
Total value	763.4	
Improvement over initial values	22%	

Source: Based on Le Roux et al. (2005).

KwaZulu-Natal that also considered an assessment of restoration), a similar assessment for most other estuaries in South Africa has not been undertaken. Taking cognizance of these impacts, and the commensurate benefits of restoring the estuarine environment resulting from dealing with the causes and consequences of this degradation, is the topic of the present work.

1.2 Study context and methodology

The present study uses estimates of total economic value (TEV, property, nursery, subsistence, and recreation) to estimate changes in TEV as a result of ecological restoration, for three estuaries in South Africa: Great Brak estuary, Knysna estuary and Swartkops estuary. The location of these estuaries is shown in Figure 1.1.

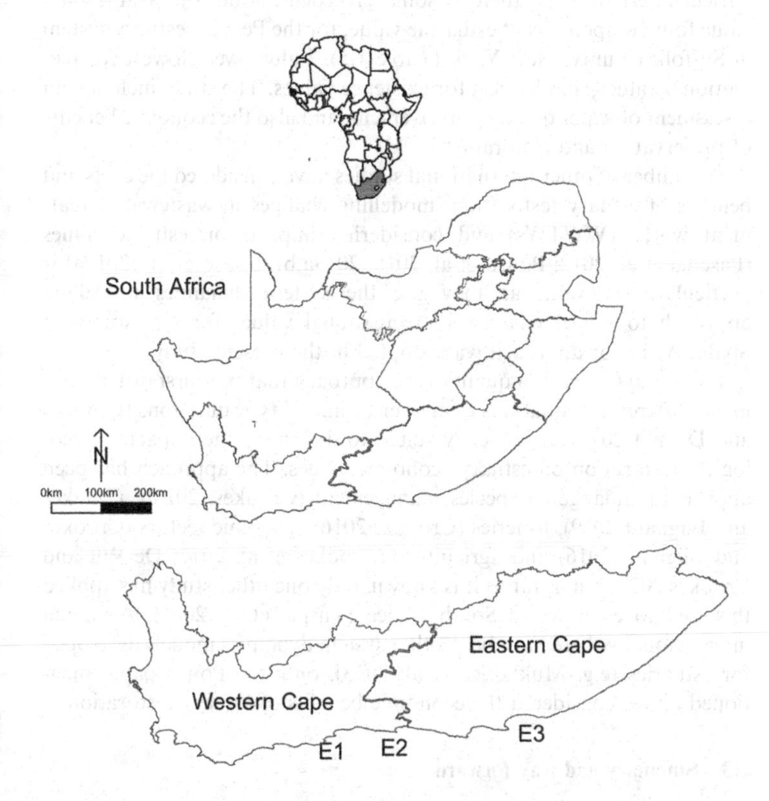

Figure 1.1 Location of the three estuaries that form part of the study. E1 = Great Brak estuary; E2 = Knysna estuary; E3 = Swartkops estuary.

Table 1.5 Estimates of the value of Peconic estuary, considering four value categories

Value	*Method*	*Value (1995 US$/acre)*
Property	Hedonic pricing model	n/a
Recreational	Travel cost model	2.41 [a]
Fishing (commercial)	Change in productivity	355.36 [b]
Existence	Contingent valuation	1,203–9,979 [c]

Notes
[a] Bird watching & wildlife viewing, Recreational fishing, Swimming, Boating
[b] Eelgrass, Saltmarsh, Inter-tidal mud flat
[c] Farmland, Undeveloped land, Wetlands, Shellfish areas, Eelgrass

Although there are no known studies to undertake this in South Africa, internationally there is some precedent. Johnston et al. (2002) value four components of estuarine value, for the Peconic estuary system in Suffolk County, New York (Table 1.5). Values were lowest for recreational values, and highest for existence values. The study included an assessment of water quality improvement, and also the economic benefits of preservation and restoration.

A number of other international studies have considered the costs and benefits of estuary restoration, modelling changes in wastewater treatment works (WWTWs) and considering impacts on estuary values (Pascual et al. 2012; Pouso et al. 2018, 2019a,b). Pouso et al. (2019a) is particularly relevant, as they use the system dynamics modelling approach to model changes in recreational values for an estuary in Spain. A similar approach was adopted in the present study.

System dynamics is a quantitative approach that uses first order, non-linear differential equations to answer "what if" type questions (Crookes and De Wit 2014). It is ideally suited to determine the impacts of ecological restoration on estuary economic values. The approach has been applied to endangered species management (Crookes, 2023b; Crookes and Blignaut 2019), fisheries (Crookes 2016) economic sectors (Crookes and Blignaut 2016) and agriculture (Crookes et al. 2017; De Wit and Crookes 2013) but as far as it is known, only one other study has applied this tool to estuaries in South Africa (Turpie et al. 2008). Although internationally there have been other system dynamics models developed for estuaries (e.g. Mukherjee et al. 2013), only the Pouso paper mentioned above considered the economic benefits of estuary restoration.

1.3 Summary and way forward

The economics of estuary restoration is an important and under-researched topic. The outline of the book is as follows. In the next chapter we consider the economic and human health impact from improved estuarine water quality in the Swartkops estuary. Chapter 3 considers the impact on ecosystem services of improved water quality. Chapter 4 examines the economic benefits of alien species removal, Chapter 5 considers the benefits on a macrophyte species resulting from the removal of an invasive plant in the Great Brak estuary, and Chapter 6 examines the economic benefits of removing invasive alien plants in the Knysna estuary.

References

Adams, J.B., Taljaard, S., Van Niekerk, L., and Lemley, D.A. 2020. "Nutrient enrichment as a threat to the ecological resilience and health of microtidal estuaries." *African Journal of Aquatic Science* 45: 23–40.

Aronson, J., Blignaut, J.N., Milton, S.J., Le Maitre, D., Esler, K.J., Limouzin, A., Fontaine, C., De Wit, M.P., Prinsloo, P., and Van Der Elst, L. 2010. "Are socioeconomic beneits of restoration adequately quantiied? A meta-analysis of recent papers (2000–2008) in restoration ecology and 12 other scientiic journals." *Restoration Ecology* 18: 143–154

Aronson, J., Milton, S.J., and Blignaut, J.N. 2007. "Restoring natural capital: Definitions and rationale." In Aronson, J., Milton, S.J. and Blignaut, J.N. (Eds). *Restoring natural capital: Business, science, and practice*. Washington D.C.: Island Press.

Balmford, A., Bruner, A., Cooper, P., Costanza, R.,. Farber, S., Green, R.E., Jenkins, M., Jefferiss, P., Jessamy, V., Madden, J., Munro, K., Myers, N., Naeem, S., Paavola, J., Rayment, M., Rosendo, S., Roughgarden, J., Trumper, K., and Turner, R.K. 2002. "Economic reasons for conserving wild nature." *Science* 297: 950–953.

Barbier, E.B., 2007. "Valuing ecosystem services as productive inputs." *Economic Policy* 22(49): 178–229.

Bark, R.H., Osgood, D.E., Colby, B.G., Katz, G., and Stromberg, J. 2009. "Habitat preservation and restoration: Do homebuyers have preferences for quality habitat?" *Ecological Economics* 68: 1465–1475

Barros, T.L., Bracewell, S.A., Mayer-Pinto, M., Dafforn, K.A., Simpson, S.L., Farrell, M., and Johnston, E.L. 2022. "Wildfires cause rapid changes to estuarine benthic habitat." *Environmental Pollution* 308: 119571.

BenDor, T.K., Livengood, A., Lester, T.W., Davis, A., and Yonavjak L. 2015. "Defining and evaluating the ecological restoration economy." *Restoration Ecology* 23(3): 209–219.

Blignaut, J.N. and Moolman, C. 2006. "Quantifying the potential of restored natural capital to alleviate poverty and help conserve nature: A case study from South Africa." *Journal for Nature Conservation* 14: 237–248.

Blignaut, J., Mander, M., Inglesi-Lotz, R., Glavan, J., and Parr, S. 2017. "Economic value of the Abu Dhabi coastal and marine ecosystem services: Estimate and management applications." In: Azar, E. and Raouf, M.A., (Eds), *Sustainability in the Gulf: Challenges and opportunities* (pp. 210–227). Abingdon: Routledge.

Brooks, M.L., D'antonio, C.M., Richardson, D.M., Grace, J.B., Keeley, J.E., DiTomaso, J.M., Hobbs, R.J., Pellant, M., and Pyke, D. 2004. "Effects of invasive alien plants on fire regimes." *BioScience* 54(7): 677–688.

Carson, R.T. and Mitchell, R.C. 1993. "The value of clean water: The public's willingness to pay for boatable, fishable, and swimmable quality water." *Water Resources Research* 29: 2445–2454

Cooper, J., Jayiya, T., Van Niekerk, L., De Wit, M., Leaner, J., and Moshe, D. 2003. An assessment of the economic values of different uses of estuaries in South Africa. CSIR Environmentek, *Stellenbosch*.

Costanza, R. 2001. "Visions, values, valuation, and the need for an ecological economics." *BioScience* 51(6): 459–468.

Costanza, R. and Cleveland, C.J. 2008. "Natural capital." In Cleveland, C.J. (Ed.). *Encyclopedia of Earth*. Washington, D.C.: Environmental Information Coalition, National Council for Science and the Environment.

Costanza, R., Wilson, M., Troy, A., Voinov, A., Liu, S., and D'Agostino, J.D. 2006. *The value of New Jersey's ecosystem services and natural capital*. Gund Institute for Ecological Economics, University of Vermont. Vermont: USA.

Crookes, D.J. 2018. "Does the construction of a desalination plant necessarily imply that water tariffs will increase? A system dynamics analysis." *Water Resources and Economics* 21: 29–39.

Crookes, D. 2023a. "Fisheries restoration: Lessons learnt from four benefit-cost models." *Frontiers in Ecology and Evolution* 11: 1067776. doi: 10.3389/fevo.2023.1067776

Crookes, D.J. 2023b. "Ecology, opportunity or threat? Drivers of the caracal (Caracal caracal) population decline in South Africa." *African Journal of Ecology* [In press] doi:10.1111/aje.13150

Crookes, D.J. and Blignaut, J.N. 2019. "Investing in natural capital and national security: A comparative review of restoration projects in South Africa." *Heliyon* 5: e01765. 10.1016/j.heliyon.2019.e01765

Crookes, D.J. and Blignaut, J.N. 2019. "An approach to determine the extinction risk of exploited populations". *Journal for Nature Conservation* 52: 125750. 10.1016/ j.jnc.2019.125750

Crookes, D.J. and Blignaut, J.N. 2016. "Predator-prey analysis using system dynamics: An application to the steel industry." *South African Journal of Economic and Management Sciences* 19(5): 733–746

Crookes, D.J. and De Wit, M.P. 2014. "Is system dynamics modelling of relevance to neoclassical economists?" *South African Journal of Economics* 82(2): 181–192

Crookes, D., Strauss, J., and Blignaut, J. 2017. "The effect of rainfall variability on sustainable wheat production in the Swartland region, South Africa." *African Journal of Agricultural and Resource Economics* 12(1): 62–84.

Crookes, D.J. 2016. "Trading on extinction: An open access deterrence model for the South African abalone fishery." *South African Journal of Science* 112(3/4): 105–113

Daly, H.E. and Cobb, J. 1989. *For the common good, redirecting the economy towards community, the environment and a sustainable future*. London: Green.

De Groot, R.S., Kumar, P., van der Ploeg, S., and Sukhdev, P. 2010. "Estimates of monetary values of ecosystem services. Appendix 3." In: Kumar, P. (Ed.). *The economics of ecosystems and biodiversity (TEEB): Ecological and economic foundations*. London: Earthscan.

De Wit, M.P. and Crookes D.J. 2013. "Improved decision-making on irrigation farming in arid zones using a system dynamics model." *South African Journal of Science* 109 (11/12): 8.

De Wit, M., van Zyl, H., Crookes, D., Blignaut, J., Jayiya, T., Goiset, V., Mahumani, B. 2012. "Including the economic value of well-functioning urban ecosystems in financial decisions: Evidence from a process in Cape Town." *Ecosystem Services* 2: 38–44.

Dorrough, J., Vesk, P.A., and Moll, J. 2008. "Integrating ecological uncertainty and farm-scale economics when planning restoration." *Journal of Applied Ecology* 45: 288–295.

Elliman, C. and Berry, N. 2007. "Protecting and restoring natural capital in New York City's watersheds to safeguard water." In Aronson, J., Milton, S.J., and Blignaut, J.N. (Eds). *Restoring natural capital: Business, science, and practice*. Washington D.C.: Island Press.

Higgins, S.I., Turpie, J.K., Costanza, R., Cowling, R.M., Le Maitre, D.C., Marais, C., and Midgley, G.F. 1997. "An ecological economic simulation model of mountain fynbos ecosystems: Dynamics, valuation and management." *Ecological Economics* 22: 155–169.

Holmes, T.P., Bergstrom, J.C., Huszarc, E., Kaskd, S.B., and Orr, F. 2004. "Contingent valuation, net marginal benefits, and the scale of riparian ecosystem restoration." *Ecological Economics* 49: 19–30.

Isley, P., Isley, E.S., and Hause, C. 2011. *Muskegon Lake Area of Concern habitat restoration project: socio-economic assessment*. Allendale, Michigan: Grand Valley State University.

Johnston R.J., Grigalunas T.A., Opaluch J.J., Mazzotta M., and Diamantedes J. 2002. "Valuing Estuarine Resource Services Using Economic and Ecological Models: The Peconic Estuary System Study." *Coastal Management* 30(1): 47–65, doi:10.1080/08920750252692616.

Kiel, K. and Zabel, J. 2001. "Estimating the economic beneits of cleaning up Superfund sites: The case of Woburn, Massachusetts." *The Journal of Real Estate Finance and Economics* 22: 163–184

Kroeger, T. 2012. *Dollars and sense: Economic benefits and impacts from two oyster reef restoration projects in the Northern Gulf of Mexico. The Nature Conservancy*. Arlington, V.A. http://www.conservationgateway.org/Files/Pages/dollars-and-sense-economi.aspx

Kruse, S.A. and Scholz, A.J. 2006. *Preliminary economic assessment of dam removal: The Klamath River*. Portland, O.R.: EcoTrust. http://www.ecotrust.org/indigenousaffairs/siskiyou_assessment.html

Le Roux, R., Nahman, A., Pillay, S., Weerts, S., and Reyers, B. 2005. "The economic impacts associated with a change in the environmental quality of the Kongweni Eastuary at Margate, KZN." CSIR, Report No. ENV-SC 2005, 51.

McCormick, B., Clement, R., Fischer, D., Lindsay, M., and Watson, R. 2010. *Measuring the economic benefits of America's Everglades restoration*. Georgia: Mather Economics, LLC, Roswell.

Millennium Ecosystem Assessment (MA). 2005. *Ecosystems and human well-being: Synthesis*. Washington D.C.: Island Press.
Mills, A.J., Turpie, J.K., Cowling, R.M., Marais, C., Kerley, G.I.H., Lechmere-Oertel, R.G., Sigwela, A.M., and Powell, M. 2007. "Assessing costs, benefits, and feasibility of restoring natural capital in subtropical thicket in South Africa." In Aronson, J., Milton, S.J., and Blignaut, J.N. (Eds). *Restoring natural capital: Business, science, and practice*. Washington D.C.: Island Press.
Milon, J.W. and Scrogin, D. 2006. "Latent preferences and valuation of wetland ecosystem restoration." *Ecological Economics* 56: 162–175.
Mukherjee, J., Ray, S., and Ghosh, P.B. 2013. "A system dynamic modelling of carbon cycle from mangrove litter to the adjacent Hoogly estuary, India." *Ecological Modelling* 252: 185–195.
Mueller, J.M., Swaffar, W., Nielsen, E., Springer, A.E., and Masek Lopez, S. 2013. "Estimating the value of watershed services following forest restoration." *Water Resources Research* 49(4): 1773–1781.
Nahman, A. 2006. *Valuing water quality changes in the Kongweni estuary, South Africa, using contingent behaviour data*. M.Sc. dissertation, University of Manchester, UK.
Nahman, A. and Rigby, D. 2008. "Valuing blue flag status and estuarine water quality in Margate, South Africa." *South African Journal of Economics* 76(4): 721–737.
NBA 2019. *South African National Biodiversity Assessment 2018: Technical Report. Volume 3: Estuarine Realm*. Pretoria: CSIR & SANBI.
Pascual, M., Borja, A., Franco, J., Burdon, D., Atkins, J.P. and Elliott, M. 2012. "What are the costs and benefits of biodiversity recovery in a highly polluted estuary?" *Water Research* 46: 205–217. doi: 10.1016/j.watres.2011.10.053
Peacock, R., Bently, M., Rees P., and Blignaut, J.N. 2023. "The benefits of ecological restoration exceed its cost in South Africa: An evidence-based approach." *Ecosystem Services* 61: 101528
Pouso, S., Borja, Á, Martín, J., and Uyarra, M.C. 2019a. "The capacity of estuary restoration to enhance ecosystem services: system dynamics modelling to simulate recreational fishing benefits." *Estuarine and Coastal Shelf Science* 217: 226–236. doi: 10.1016/j.ecss.2018.11.026
Pouso, S., Ferrini, S., Turner, R.K., Borja, A., and Uyarra, C.M. 2019b. "Monetary valuation of recreational fishing in a restored estuary and implications forfuture management measures." *ICES Journal of Marine Science* 9: fsz091. doi: 10.1093/icesjms/fsz091
Pouso, S., Ferrini, S., Turner, R.K., Uyarra, M.C., and Borja, Á. 2018. "Financial inputs for ecosystem service outputs: beach recreation recovery after investments in ecological restoration." *Frontiers in Marine Science* 5: 375. doi: 10.3389/fmars.2018.00375
Tong, C., Feagin, R.A., Lu, J., Zhang, X., Zhu, X., Wang, W., and He, W. 2007. "Ecosystem service values and restoration in the urban Sanyang wetland of Wenzhou, China." *Ecological Engineering* 29: 249–258.
Turpie, J.K., Forsythe, K.J., Knowles, A., Blignaut, J., and Letley, G. 2017. "Mapping and valuation of South Africa's ecosystem services: A local perspective." *Ecosystem Services* 27: 179–192.

Turpie, J.K., Clark, B.M., Cowley, P., Bornman, T., and Terörde, A. 2008. *Integrated ecological-economic modelling as an estuarine management tool: a case study of the East Kleinemonde Estuary. Volume II*. Pretoria: Model construction, evaluation and user manual. WRC Report No. 1679/2/08.

Turpie, J., Lannas, K., Scovronick, N., and Louw A. 2010. "Wetland ecosystem services and their valuation: A review of current understanding and practice", WRC Report No. TT 440/09

Valderrama, A., Levine, L., Bloomgarden, E., Bayon, R., Washowicz, K., Kaiser, C., Holland, C., Ranney, N., Scott, J., Kerr, O., DePhilip, M., Davis, P., Devine, J., Garrison, N., and Hammer, R. 2013. "Creating clean water cash flows: developing private markets for green stormwater infrastructure in Philadelphia." R:13-01-A. Natural Resources Defense Council. January 2013. (accessed 21 Mar 2015)

Van Niekerk, L., Adams, J.B., Lamberth, S.J., MacKay, C.F., Taljaard, S., Turpie, J.K., Weerts, S.P., and Raimondo, D.C. 2019 (eds). *South African National Biodiversity Assessment 2018: Technical Report. Volume 3: Estuarine Realm*. CSIR rechport number CSIR/SPLA/EM/EXP/2019/0062/A. South African National Biodiversity Institute, Pretoria. Report Number: SANBI/NAT/NBA2018/2019/Vol3/A. http://hdl.handle.net/20.500.12143/6373

Vickerman, S. 2013. *Nature's benefits: The importance of addressing biodiversity in ecosystem service programs*. Washington D.C.: Defenders of Wildlife, http://www.defenders.org/publication/natures-benefits-importance-addressing-biodiversity-ecosystem-service-programs (accessed 21 March 2015)

Vundla, T., Blignaut, J.N., and Crookes, D. 2017. "Financing active restoration in South Africa: An evaluation of different institutional models." *African Journal of Agricultural and Resource Economics* 12 (4): 430–453.

Weinerman, M., Buckley, M., Reich, S. 2012. *Socioeconomic benefits of the Fischer Slough restoration project*. Portland, O.R.: ECONorthwest.

2 The recreational and human health costs from declining water quality in the Swartkops estuary, South Africa

2.1 Introduction

The economic impact of land-based pollution sources to estuaries and coastal waters is an important but as of yet under-investigated research topic in the South African context. The National Biodiversity Assessment (2019) indicated that our estuaries are under severe pollution pressure and that improvement of water quality as a key intervention would lead to substantial improvement in estuary health and associated benefits that society derives from them (Van Niekerk et al. 2019). Restoration of coastal water quality is investigated at the Swartkops estuary as it is impacted by industrial discharges, inputs from three wastewater treatment works (WWTWs), stormwater runoff and agricultural return flow. The benefits of water quality improvement on estuary health and the delivery of ecosystem services are assessed, and inputs are made to the development of a socio-ecological systems framework for the restoration of estuaries.

2.2 Methodology

The following socio-ecological systems framework is proposed (Figure 2.1). It is evident from this that pollutants impact two aspects of ecosystem services in the Swartkops estuary: 1) estuary health and 2) human well-being. In this chapter, we consider the economic benefits of improving water quality on both of these elements.

A list of pollutants and their impact on estuary and human health are given in Table 2.1. Table 2.1 indicates that these pollutants both cause fish mortality and have human health impacts. Therefore, it is important to value both of these effects.

Figure 2.2 indicates the relevant economic valuation techniques applicable to this study. First, an improvement in environmental quality (through ecological restoration) improves estuarine health. This has an impact on estuarine fish stocks. The change in productivity approach is

DOI: 10.4324/9781032651675-2

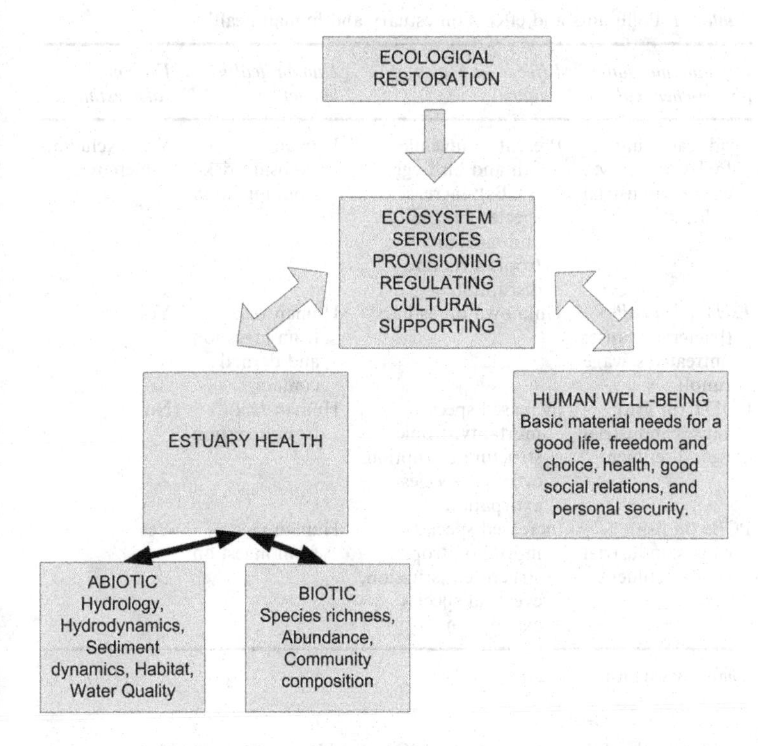

Figure 2.1 Socioecological systems framework for the Swartkops estuary.
Source: Adapted from Adams et al. (2020).

used when a change in environmental quality affects the productivity of the resource. Usually, a dose response relationship is required to relate the change in environmental quality to the effects on the productivity of the resource, but these relationships may be difficult to estimate in practice. A solution proposed by Blignaut and Lumby (2004) is through the use of sensitivity analysis.

A second impact of improvements in environmental quality as a result of restoration is the human health effects (Figure 2.2). Human health benefits include both the reduction in the morbidity, but also the reduction in premature mortality from improvements in estuarine water quality. The cost of illness approach measures sickness-related costs, such as medicine, doctor's visits and hospitalization, while the human capital approach measures the associated effects on the productivity of labour (De Wit et al. 2009).

Table 2.1 Pollutants and effects on estuary and human health

Organic, inorganic, physio-chemical	*Effects on health of estuary*	*Human health impact*	*Economic value estimated*
Lead, cadmium, PCBs, mercury; causes: industrial effluent	Present in juvenile fish and bird eggs (PCBs); increase species morbidity and mortality, tropic structure disruption	Human exposure risk from ingestion	Yes (excluding mercury)
Escherichia coli (bacteria); cause: untreated sewage runoff	Unknown on estuary	Human risk from ingestion and dermal contact	Yes
t-DDT (in fish); causes: industrial runoff, effluent	increased species mortality, tropic structure disruption, eventual species extirpation	Human risk from ingestion	No
PCBs (in fish); causes: industrial runoff, effluent	increased species mortality, tropic structure disruption, eventual species extirpation	Human risk from ingestion	Yes

Source: Based on Moore (2017).

The methodology for valuing COI and HC is similar to the change in productivity: Data are needed on: 1) the amount of pollutant (e.g. tonnes per annum); 2) the dose response relationship (effect of a unit change in pollutant on human health); and 3) the market value of health impact (loss of earnings, cost of illness) (De Wit et al. 2009). Good data are available on dose response relationships for human health and the cost of illness in both the US and Europe. How these values are transferred to a developing country context such as South Africa is not without controversy. Some argue that adjustments should be made to account for the differences in the standard of living between the two countries (Winpenny 1995). Normally, this is an acceptable approach when benefits transfer studies are utilized. For example, if an international economic value of an environmental good is transferred to a local context, then appropriate adjustments for purchasing power between the source and destination country would be appropriate (Van Zyl and De Wit 2013). It is important to note that these valuation techniques are not placing a value on human life (De Wit et al. 2009). However, in this case adjusting for purchasing power parity (PPP) has been suggested by some

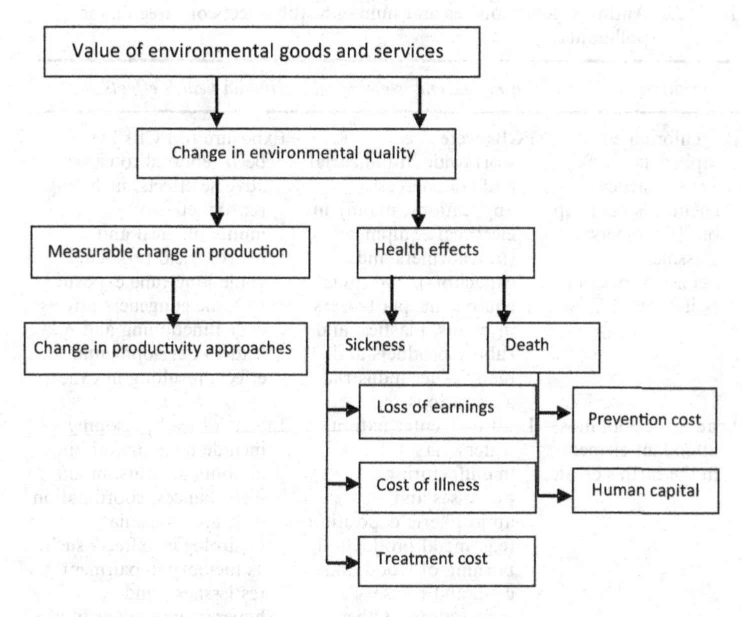

Figure 2.2 Economic valuation techniques for assessing a change in environmental quality.

Source: Adapted from Blignaut and Lumby 2004.

to indicate that a person in a developing country is worth less than a person in a developed country. In order to overcome these problems, values are estimated both without adjustments for PPP as well as with PPP adjustments, in order to obtain a range of values.

The human health effects of pollutants in the Swartkops estuary are estimated for the following pollutants: 1) PCBs; 2) lead and 3) cadmium. The reason why these three pollutants were selected is that they are known to be present in fish (which are consumed by users of the Swartkops estuary on a regular basis) and therefore could have human health implications for these users and their families. Table 2.2 indicates that these three pollutants are found in various industrial processes, domestic wastewater and other anthropogenic activities, and that human health impacts include immunological and neurological impacts (PCBs), nausea, vomiting and abdominal pains (lead poisoning), and lung disease and kidney failure (cadmium). The methodology for estimating these human health impacts for each pollutant is given below.

Table 2.2 Anthropogenic sources and human health effects of three major pollutants

Description	*Anthropogenic sources*	*Human health effects*
Polychlorinated biphenyls (PCBs) are synthetic chemicals made up of 209 isomers classified as persistent organic pollutants (POPs).	PCBs were used worldwide in industrial and commercial applications, mainly in electrical equipment (transformers and capacitors), hydraulic equipment, plasticizers in paints, plastics, and rubber products and many other industrial applications.	Exposure to PCBs has been reported to cause adverse effects, including reproductive, immunological and neurological problems, while long-time exposure to some congeners affects liver functioning and may lead to developmental effects resulting in cancer.
Lead is the 36th most abundant element in the earth's crust.	Lead may enter natural waters via manufacturing processes and atmospheric deposition (e.g. metal production, burning of wood and coal, and refuse incineration). Other sources include domestic wastewater and sewage.	Effects of lead poisoning include nausea, vomiting, abdominal pains, mood disturbances, coordination loss, and anaemia. Neurological effects such as memory impairment, restlessness, and hyperactivity are examples of more severe situations. Exposure to lead may also lead to miscarriages for pregnant women.
Cadmium is a toxic metal that exists as complex oxides, sulphides, and carbonates in zinc, lead, and copper ores.	Sources of cadmium are metallurgical industries, municipal effluents, sewage sludge, pigments and plastics.	Profound exposures to Cd may cause severe respiratory irritation, while occupational exposures may lead to chronic lung diseases or testicular degeneration. Lower concentrations of Cd exposure may damage the functional units of the kidney, resulting in kidney damage or failure. Cadmium also affects the loss of calcium that can lead to the weakening of the bones, generally known as the Itai-Itai disease.

Sources: Kampire et al. (2015a, b); Nel (2014).

2.2.1 *Estuary health*

The economic costs of declining water quality in the Swartkops estuary are first the loss in the value of fish productivity. Table 2.1 indicated that the effects of pollutants such as PCBs, lead and cadmium could result in fish mortality. Therefore, there is likely to be a decline in fish productivity as a result of pollutants in the estuary. The annual fishing value of Swartkops estuary is reported in Cooper et al. (2003) as R55.7 million (2002 prices). These values were then adjusted to 2019 prices using the Consumer Price Index (2019). This gives a value of R136.2 million per annum (2019 prices). Since we do not know the effect of a change in water quality on fish productivity, a range of estimates were used: 10% decline in productivity (low estimate); 20% decline in productivity (high estimate). This is a loss in recreational benefits as a result of poor water quality.

2.2.2 *Human health impacts*

We utilize estimates of concentration of PCBs, lead and cadmium in three fish species commonly consumed by users of the Swarkops estuary (P. commersonnii, L. amia and A. Japonicus), as reported in Nel et al. (2015), to determine the amount of pollutant that users are exposed to. We use a range of concentrations, a low estimate and a high estimate, based on the minimum and maximum concentration levels in Nel et al. (2015). The number of users of the Swarkops estuary (number of households using) is given in Table 2.3. Estimates of the number of people who consume fish at

Table 2.3 A summary of the mean number of anglers using Swartkops estuary over weekends and a comparison of their methodology and approach.

	Mean number of anglers per weekend day[1]	*Days fished*[2]	*Days fished per annum*[3]	*Number of anglers (Swartkops)*		
		Min	*Max*	*Min*	*Max*	
Shore anglers	**117.5**	**16.25**	**17.8**	**1910**	**2092**	**(A)**
Recreational anglers	90.5				1611	77%
Subsistence anglers	27.0				481	23%
Skiboat anglers	**30.5**			**823**	**823**	**(B)**
Recreational	30.5		27		823	
Commercial	n/a					
Total				**2733**	**2915**	**(A)+(B)**

Notes
1 Baird et al. (1996)
2 Lamberth & Turpie 2003 (East coast)
3 McGrath et al. 1997 (National)

the Swartkops estuary is difficult to obtain. For example, Hosking et al. (2004) estimated that around 2,500 anglers utilize the Swartkops estuary for fishing, but it is not clear if this includes boat-based anglers and subsistence fisherfolk. Table 2.3, using data from different sources, also indicates that anglers (including subsistence and skiboat) could range between 2,733 and 2,915 anglers. Assuming six persons per household, this means that a maximum number of people consuming fish in the Swartkops estuary is around 42,193 (1996 data). This equated to 4% of the Nelson Mandela Bay (NMB) municipality in 1996. Assuming that the number of users grew in proportion to the NMB population, this implies a total number of consumers of fish from the Swartkops estuary in 2019 of 55,749 people. We assume that these anglers are regular users of the Swartkops estuary, and therefore that consumption of fish from the Swartkops estuary results in long-term exposure to these pollutants.

Although PCBs have the most serious consequences, potentially causing cancer at high doses (Table 2.2), concentrations of PCBs in fish in the Swartkops estuary are still relatively low, the highest being in A. Japonicus at 0.086 ppm (Nel et al. 2015). By contrast, dose response relationships indicate that PCB concentrations that cause cancer typically range from 1 to 5 ppm (Gold and van Ravenswaay 1984). However, we extrapolate the dose response relationships to the lower concentration levels by using polynomial interpolation. The study of Gold and Van Ravenswaay (1984) also reports cost of illness and foregone earnings associated with PCBs, which were then adjusted using estimates in Ofiara and Brown (1999), and then converted into 2019 Rands.

Marais (1988) estimated that the annual catch of anglers at the Swartkops estuary is 11,151 kilograms. Using data in Nel et al. (2015) it is possible to estimate the quantity of cadmium in fish caught by anglers. We can then multiply this by the health cost of cadmium in Denmark (Euro 334/kg Cd, 2010 prices), as reported by Pizzol et al. (2014), and convert to 2019 prices.

For lead contamination, Levin (2016) estimate that 10,000 US workers exposed to long-term lead contamination had a health-rated damage cost of $392 million (2014 prices), based on a shift in lead concentrations of 20ug/dl. Lead concentrations reported in Nel et al. (2015) are actually on the high side (averaging between 38 and 53 ug/dl across the three fish species). We use the ratio of US workers to the US population, and multiply by the number of NMB residents to determine the number of people likely to be exposed to high levels of lead concentrations.

2.3 Results

Roughly two thirds of the economic cost from declining water quality (low scenario) and one third of economic cost (high scenario) in Swartkops

Table 2.4 Economic costs associated with declining water quality at Swartkops estuary (2019 R million/year)

	Value (cost)		*Comment*
	Low	*High*	
Estuarine health improvement			
Declining fish productivity	13.6	27.2	Low = 10% loss; High = 20% loss
Declines in bait species and bird populations	n/a	n/a	Impacts on bait species health and bird populations has an additional cost for anglers and bird watchers. Not valued.
Human health costs	6.02	71.10	Value of declines in human health (reduced loss of earnings, cost of illness) resulting from the consumption of contaminated fish
PCBs	0.1%	0.3%	Proportion of human health cost that relates to PCBs
Cd	3.9%	40.4%	Proportion of human health cost that relates to cadmium
Pb	96%	59.3%	Proportion of human health cost that relates to lead
Escherichia coli (bacteria)	0.56	2.90	Cost of illness
t-DDT (in fish)	n/a	n/a	Not valued
Costs associated with other recreational activities (boating, walking, picnicking, swimming)	n/a	n/a	Pollutants may emit an unpleasant smell or cause eutrophication, and could result in an declining amenity value for recreational users. Not valued
Total (2019 R million)	>20.2	>101.2	
Annualize value (US$/ha/year)	1,362	6,823	2020 prices

estuary is from declining fish productivity, and the other one third (two thirds = high) is from increased human health impacts (Table 2.4).

It is worth noting, though, that this is not an exhaustive assessment of the costs of pollutants. Only three are focused on here, namely PCBs, Cd and Pb. In terms of the human health impacts from declining water quality.

PCBs has a relatively small cost (between 0.1% and 0.3% of human health cost), whereas high lead concentrations has the greatest cost (between 59.3% and 96% of human health cost) (Table 2.4).

It is worth emphasizing that many values are omitted from the study, such as other recreational costs from declining water quality, as well as the value on bait species. Even so, the cost of impacts in the Swartkops estuary could amount to hundreds of millions of Rands (Table 2.4).

These costs need to be compared with the costs of restoring the Swartkops estuary. This was not undertaken in the present chapter, but a case study for the Sezela estuary on the KwaZulu-Natal South Coast reported in Claassens et al. (2020) indicates the potential range of values. The estuary was adversely impacted from pollutants from a nearby sugar mill that impacted fish populations dramatically. When the sugar mill improved their processing of effluent, the estuarine water quality improved, and fish biota returned to the estuary.

2.4 Discussion

The economic costs from declining water quality at Swartkops estuary, on the basis of declining fish productivity and increased human health cost amounts to a loss in value of the estuary of between 15% and 74% (based on Cooper et al. 2003's estimate of the value of Swartkops estuary). It is important to note that these costs (fish productivity losses and cost of human health) as a result of estuary degradation are not included in the estimates of value reported in the literature. This would necessarily decrease the value of estuaries in their degraded state. Estimating human health impacts is controversial, but nonetheless important.

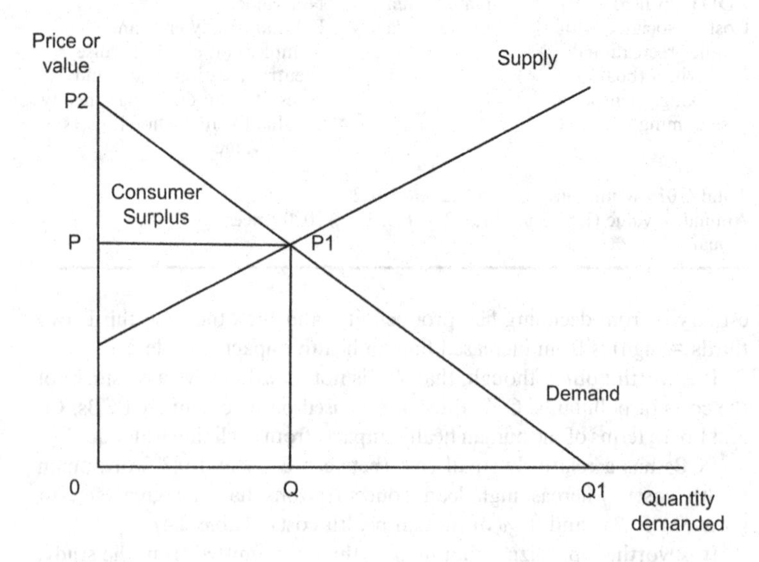

Figure 2.3 Supply and demand curve for an environmental good or service.

It is also important to note that estimates of health impacts (cost of illness and human capital approaches) measure actual expenditure, and the values are typically lower than those that measure value based on the area whole demand curve (Figure 2.3). In the Figure, estimates based on spending (cost of illness, human capital), are reflected in the area P-P1-0-Q. Whereas, estimates based on stated preferences (contingent valuation) and revealed preferences (travel cost method) estimate value based on the whole area under the demand curve (P2-Q1-0) (see Winpenny 1995 for a further elaboration). The values reported on in this chapter are therefore to be regarded as lower bound estimates, as they do not capture the Total Economic Value (TEV) but only market prices.

Note also that only two impacts are captured, (partial) recreational values and (partial) human health costs.

References

Adams, J.B., Whitfield, A.K., and Van Niekerk, L. 2020. "A socio-ecological systems approach towards future research for the restoration, conservation and management of southern African estuaries." *African Journal of Aquatic Science* 45(1–2): 231–241.

Blignaut, J.N. and Lumby, A. 2004. "Economic valuation." In Blignaut, J.N. and de Wit, M.P. (Eds). *Sustainable options: Economic development lessons from applied environmental and resource economics in South Africa.* Cape Town: UCT Press.

Cooper, J., Jayiya, T., Van Niekerk, L., De Wit, M., Leaner, J., and Moshe, D. 2003. *An assessment of the economic values of different uses of estuaries in South Africa.* Stellenbosch: CSIR Environmentek.

De Wit, M.P., Van Zyl, H., Crookes, D.J., Blignaut, J.N., Jayiya, T., Goiset, V., and Mahumani, B.K. 2009. "Investing in natural assets: A business case for the environment in the city of Cape Town." Report prepared for the City of Cape Town, Cape Town, South Africa, 18 August.

Gold, M.S. and Ravenswaay, E.O.V. 1984. *Methods for assessing the economic benefits of food safety regulations: A case study of PCBs in fish.* Agricultural Economics Report No. 460. East Lansing: Department of Agricultural Economics, Michigan State University.

Kampire, E., Rubidge, G. and Adams, J.B. 2015a. "Distribution of polychlorinated biphenyl residues in several tissues of fish from the North End Lake, Port Elizabeth, South Africa." *Water SA* 41(4): 559–570.

Kampire, E., Rubidge, G. and Adams, J.B. 2015b. "Distribution of polychlorinated biphenyl residues in sediments and blue mussels (Mytilus galloprovincialis) from Port Elizabeth Harbour, South Africa." *Marine Pollution Bulletin* 91(1): 173–179.

Levin, R. 2016. "The attributable annual health costs of US occupational lead poisoning." *International Journal of Occupational and Environmental Health* 22(2): 107–120.

Marais, J.F.K. 1988. "The ichthyofauna of the Swartkops Estuary." In: Baird, D., Marais, J.F.K. and Martin, A.P. (Eds.). *The Swartkops Estuary: Proceedings of a symposium held on 14 and 15 September 1987 at the University of Port Elizabeth*. South African National Scientific Programmes Report No. 156. pp. 76–85.

Moore, G.D. 2017. *Swartkops Estuary: Port Elizabeth, South Africa*. Johns Hopkins Whiting School of Engineering. 22 April 2017.

Nel, L., Strydom, N.A. and Bouwman, H. 2015. "Preliminary assessment of contaminants in the sediment and organisms of the Swartkops Estuary, South Africa." *Marine pollution bulletin* 101(2): 878–885.

Nel, L. 2014. *Presence, levels and distribution of pollutants in the estuarine food web: Swartkops River Estuary, South Africa*. M.Sc thesis, Potchefstroom: North-West University.

NBA 2019. *National Biodiversity Assessment 2019*. Pretoria: SANBI.

Ofiara, D.D. and Brown, B. 1999. "Assessment of economic losses to recreational activities from 1988 marine pollution events and assessment of economic losses from long-term contamination of fish within the New York Bight to New Jersey." *Marine Pollution Bulletin* 38(11): 990–1004.

Pizzol, M., Smart, J.C. and Thomsen, M. 2014. "External costs of cadmium emissions to soil: a drawback of phosphorus fertilizers." *Journal of Cleaner Production* 84: 475–483.

Van Niekerk, L., Adams, J.B., Lamberth, S.J., MacKay, C.F., Taljaard, S., Turpie, J.K., Weerts, S.P. and Raimondo, D.C., 2019 (eds). *South African National Biodiversity Assessment 2018: Technical Report. Volume 3: Estuarine Realm*. CSIR report number CSIR/SPLA/EM/EXP/2019/0062/A. Pretoria: South African National Biodiversity Institute. Report Number: SANBI/NAT/NBA2018/2019/Vol3/A.

Van Zyl, H. and De Wit, M.P. 2013. *Environmental impact assessmemt (EIA) for the proposed N3: Keeversfontein to Warden (De Beers Pass Section)*. DEA ref. no. 12/12/20/1992. Environmental Resource Economics DRAFT Specialist Report.

Winpenny J.T. 1995. *The economic appraisal of environmental projects and policies – A practical guide*. Paris, France: OECD & ODI.

3 Improving water quality in the Swartkops estuary

Impacts on ecosystem services

3.1 Introduction

The impact of land-based pollution sources to estuaries and coastal waters is a global problem. Accelerated anthropogenic nutrient loading, particularly in the form of nitrogen (N) and phosphorus (P), is widely accepted as the key culprit responsible for widespread cultural eutrophication in estuarine and coastal waters. It is estimated that approximately 24% of land-based anthropogenic N inputs to coastal watersheds reach coastal waters (Malone and Newton 2020). Brondizio et al. (2019) estimated that over 80% of urban and industries wastewater is released into river systems without adequate treatment. The solution is the ecological restoration of coastal systems, rivers and wetlands (Sachs et al. 2019). Pollution of estuaries due to land-based activities is significant in the South African context. The National Biodiversity Assessment of 2018 indicated that South Africa's estuaries are under severe pollution pressure (Adams et al. 2020; Taljaard et al. 2017; Van Niekerk et al. 2019). Water pollution and the degradation of river systems incurs significant cost in terms of the value of ecosystem services lost (Evans 2020). Improvement of water quality and riparian area restoration is therefore a key intervention that would lead to significant improvement in estuary health and associated benefits that society derives from them. Although much is known about the biotic and abiotic conditions of estuaries, relatively fewer studies have focused on the restoration of estuaries (Claassens et al.) and even fewer on the economic benefits thereof in terms of ecosystem services improvements (e.g. property, nursery, subsistence and tourism values) (Adams et al. 2020).

The ecological functioning of the Swartkops estuary has been significantly altered in recent decades, with the system now presiding in a permanent eutrophic condition (Adams et al., 2019; Lemley et al. 2017). A study by Lemley et al. (2019) estimated that, since 2013, three upstream wastewater treatment works (WWTWs) (i.e. Kelvin Jones,

DOI: 10.4324/9781032651675-3

Despatch and Kwanobuhle) discharge combined annual inorganic nutrient loads of 1.5 ×10x 10^5 kg N and 4.2 ×10x 10^4 kg P to the Swartkops estuary. The occurrence of high-biomass harmful algal blooms (HAB) and bottom-water oxygen depletion (< 2 mg l^{-1}) in the stratified middle to upper reaches of the estuary are the most notable eutrophication symptoms culminating from these excessive anthropogenic nutrient inputs. To date, the most common HAB-forming species in the estuary belong to the Dinophyceae (*Peridinium* sp.; Lemley et al. 2017) and Raphidophyceae (*Heterosigma akashiwo*; unpublished data) phytoplankton classes, occasionally forming dense accumulations (> 100 μg Chl-*a* l^{-1}) that discolour estuarine waters and influence the dynamics of faunal communities (e.g. fish, invertebrates). Interestingly, the high P loads introduced via the upstream WWTWs often result in heavily skewed inorganic N:P ratios (≤ 1:1). As such, the implementation of dual-nutrient reduction strategies (Freeman et al. 2019) and diversion of point source discharges for reuse (e.g. artificial wetlands, non-consumptive irrigation purposes) are considered key to the effective management of the Swartkops estuary into the future.

This chapter focuses on evaluating the benefits of improving water quality on a heavily degraded estuary in South Africa. It evaluates whether it is more economically viable to reduce the nutrient loads in the river system or decrease the flow of effluent from the WWTWs, or a combination of the two. A cost-benefit analysis is conducted by developing a system dynamics model. In the next section, the study site is described.

3.2 Study site

Swartkops estuary, an ecologically important estuary in South Africa, is heavily degraded (Van Niekerk et al. 2019). A major cause for concern is the high levels of eutrophication due to heavy nutrient loads (particularly dissolved inorganic nitrogen and phosphorous) emerging from WWTWs in the area that have impacted on the ecological functioning of the estuary (Adams et al. 2019; Lemley et al. 2019; Snow et al. 2019).

The Swartkops estuary is situated in the Eastern Cape province with the mouth of the estuary near the city of Port Elizabeth (Figure 3.1). The Swartkops river catchment is situated within primary catchment M (Water Management Area 15), and includes the M10A (KwaZunga), M10B (Elandsrivier), M10C (Swartkops River: Elands Confluence), and M10D (Swartkops River: Despatch – River Mouth) quaternary catchments (Enviro-Fish Africa 2009). WWTWs within the catchment are a major source of phosphorous, nitrogen and cholorphyll, which have resulted in the growth of water hyacinth (*Eichhornia crassipes*) and other invasive aquatic plants in the river (Adams et al. 2019).

Figure 3.1 Map showing the location of Swartkops estuary.

The main source of nutrients in the river comes from the Kelvin Jones WWTW (Adams and Riddin 2020), but other WWTWs discharging to the river system (Kwanobuhle and Despatch) are also important nutrient sources. Fifty percent of the nitrogen and phosphorous in the river system comes from these three WWTWs (Adams and Riddin 2020). The Motherwell canal is also a source of nutrients, particularly inorganic N, but an artificial wetland has already been constructed at this site with the management objective of reducing nutrient and bacterial loads. However, preliminary results from ongoing research suggest that the small size of the wetland and little to no harvesting of wetland plants (i.e. *Typha capensis*), have culminated in a system largely ineffective at reducing incoming nutrient loads. Instead, the flow volumes entering the estuary from the wetland outlet have been shown to be equivalent to inputs from a medium-sized WWTW (i.e. 5–9 Ml d^{-1}), with the discharges introducing significant inorganic N loads (particularly nitrate) and elevated P loads (compared to input loads) to the Swartkops estuary. Other notable nutrient point sources to the estuary include the industrially-influenced Markman Canal and the Chatty River that is subject to diffuse inputs from informal settlements (Adams et al. 2019).

The present study focuses exclusively on riparian area rehabilitation, and also on mitigating the effects of nutrient load pollution from these three WWTWs, either through rehabilitating the WWTWs or alternatively by constructing artificial wetlands. The costs of each of these two interventions are compared to find the most economically viable alternative.

3.3 Methodology

3.3.1 Socioecological systems framework

Social ecological systems approaches are important for modelling ecosystem processes incorporating both estuarine health, as well as human welfare components (Adams et al. 2020). Ecological restoration, as we have seen earlier, is important for creating jobs, improving human health, increasing the value of ecosystems and ecological infrastructure, and promoting conservation (Andersson et al. 2019); Reyes-García et al. 2019). Ecological restoration is also an important mechanism to improve ecosystem health (Young 2000). There are a number of remediation research projects in the Swartkops (treatment trains, salt pans rehab/ development, artificial wetlands) (NMU 2020). Swartkops is a highly important but degraded estuary, which is impacting human health and its ability to function in terms of providing ecosystem services and ecological functioning (e.g. Adams et al. 2019; Olisah et al. 2019, 2020; van Aswegen et al. 2019). Systems approaches and scenarios are important tools for modelling systems such as these that are in need of ecological restoration (Zweig et al. 2020). The purpose of this chapter is to assess the costs and benefits of restoration on the estuarine health (water quality), using a social-ecological systems approach on the Swartkops estuary (Basconi et al. 2020).

3.3.2 Conceptual model

The following socio-ecological systems framework is proposed for the Swartkops Estuary Restoration (SER) Model (Figure 3.2). The river is highly polluted by nutrients (phosphates and nitrates). The objective is to improve the water quality of the river inflow to the estuary and thereby to assess the economic benefits of such improvements. The approach adopted is to compare the benefits in terms of the value of ecosystem improvements as a result of an improvement in water quality (resulting from a reduction in nutrient flows from the WWTWs and

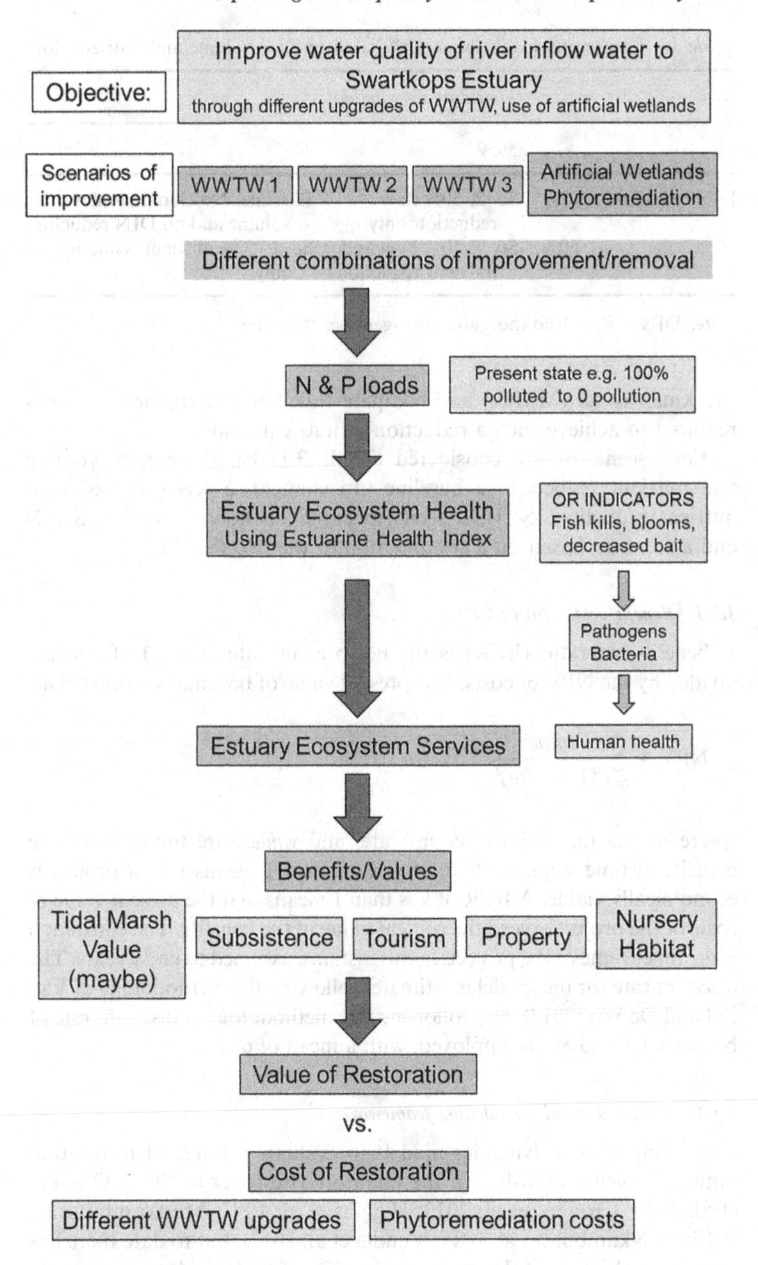

Figure 3.2 Conceptual model for the SER model.

Source: Janine Adams.

Table 3.1 Four scenarios in the model, based on flow volume and nutrient flow

		Flow Volume (m3/day)	
		50%	*100%*
DIN reduction	0%	Sc. 2. 50% flow reduction only	Baseline. No change in flow volume and no DIN reduction
	50%	Sc. 3. 50% flow and nutrient reduction	Sc. 1. 50% nutrient reduction only

Note: DIN = dissolved inorganic nitrogen; Sc. = Scenario

Markman canal (MMC)) and compare this with the engineering costs required to achieve such a reduction in nutrient loads.

Four scenarios are considered (Table 3.1) based on flow volume and nutrient reduction: a baseline (no change), a scenario based on nutrient reduction (Sc.1), a scenario based on flow reduction (Sc. 2) and a scenario based on a combination of the two (Sc. 3).

3.3.3 *Benefit-cost ratio (BCR)*

A benefit-cost ratio (BCR) is the net present value (NPV) of benefits divided by the NPV of costs. The present value of benefits is estimated as:

$$NPV = \sum_{i=1}^{n} \frac{values_i}{(1 = rate)^i}$$

where *rate* is the social discount rate, and $values_i$ are the costs or the benefits in time step*i*. A BCR of greater than 1 means that a project is economically viable. A BCR of less than 1 means that the present value of costs of the project exceed the present value of the benefits. The constant *n* is the time frame of the project, in this instance assumed to be 30 years. The discount rate for the model is estimates following the methodology of Van Zyl and De Wit (2013), and following this methodology, a discount rate of between 4% and 8% is employed, with a mean of 6%.

3.3.4 *System dynamics modelling framework*

Employing system dynamics models to evaluate a range of restoration options is well established in the literature (Bester et al. 2019; Crookes et al. 2020; Crookes et al. 2013; Higgins et al. 1997; Mudavanhu et al. 2017a,b; Nkambule et al. 2017; Vundla et al. 2017), but to date there has been no evaluation of the economic benefits of ecological restoration for the Swartkops estuary using system dynamics and cost-benefit analysis.

The 2018 NBA for estuaries indicated that the Swartkops estuary was a category D estuary (60%–41% of pristine, average 50% degraded) and that the following restoration activities were required (Van Niekerk et al. 2019):

- Restore/protect base flows
- Restore floods
- Improve (urban) river water quality
- Rehabilitate riparian areas/wetlands
- Remove alien vegetation
- Control recreational activities impacting birds
- Remove/reduce fishing pressure/bait collection
- Investigate eradication of alien fish

In this case, a system dynamics model is developed to evaluate two different restoration options—1) capital and 2) operating cost of improvements to the Kelvin Jones, Kwanobuhle and Dispatch WWTWs—based on the four scenarios in order to evaluate which is the most cost effective option.

The stock flow diagram of the SER model is given in Figure 3.3.

The input data used in the model are given in Appendix 1. Ecosystem values due to the estuary include tourism, nursery, property values and subsistence benefits, and are specific to the Swartkops estuary (Turpie et al. 2017). The ecosystem values are linked to the present ecological condition of the estuary based on the classification scheme given in Table 3.2.

3.4 Results

3.4.1 Wastewater treatment capital cost

Capital cost of removing nutrients from wastewater flows is given in Figure 3.4. This study and the Northern WWTW assume activated sludge and biological nutrient removal (BNR), whereas the Motetemba WWTW uses algal pond technology. The choice of technology is informed by the flow volume. The total flow from the three WWTWs and the MMC total 34,110 m3/day, and as a result a macro WWTP (>25 Ml/d) would be needed, following the definitions of Oberholster (2020). Operating costs for BNR and activated sludge range between 1.0% (this study) and 3.7% (Mitchell et al. 2014).

3.4.2 Economic value of estuaries

Estimates of the value of different estuaries and their ecological state are given in Table 3.3. The values are in 2020 Rand per hectare.

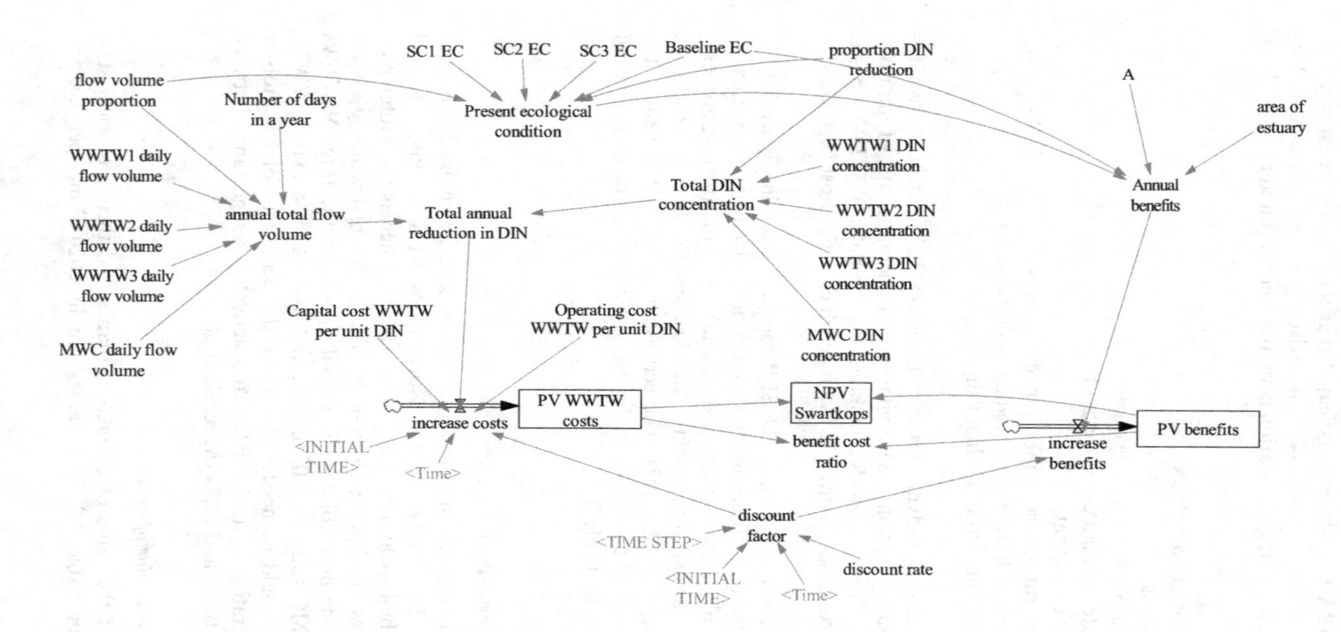

Figure 3.3 Stock flow diagram for the SER model.

Table 3.2 Classification of present ecological condition

Excellent	1	Near-natural conditions with no anthropogenic induced stresses on estuarine biota
Good	2	No excessive primary production or organic loading, with acceptable oxygen/pH levels with no marked stress on estuarine biota
Fair	3	Moderate primary production and organic loading with some detrioration in oxygen/pH conditions causing moderate stress in fauna
Poor	4	Significant primary production and organic loading resulting in hypoxia and some pH variability causing significant stress on estuarine biota
Degraded	5	High primary production and organic loading resulting in
Degraded	6	hypoxia/anoxia and high pH variability causing marked disfunctionality in estuarine biota

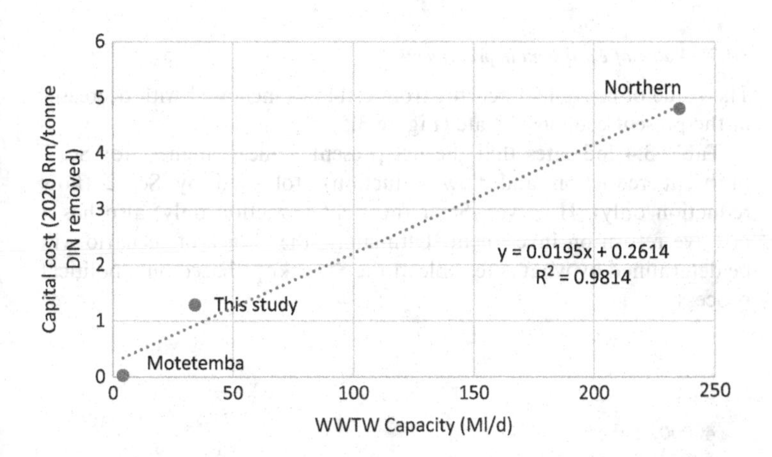

Figure 3.4 Capital cost of removing nutrients (2020 Rm/tonne DIN removed).

Source: Motetemba WWTW: Oberholster (2020); Northern WWTW (Mitchell et al. 2014); This study. Based on nutrients flows at Swartkops.

The ecosystem services valued included: 1) Subsistence value, 2) Property value, 3) Tourism value and 4) Nursery value. The highest ecosystem value was for the Kowie estuary, at R5.5 million/ha. The lowest ecosystem value was for the Swartkops estuary, at R0.5 million/ha. Swartkops estuary is also the most ecologically degraded of the nine estuaries considered as part of this analysis, with a present ecological state of 4 (Table 3.3).

Table 3.3 Characteristics of nine estuaries and their ecological state

Estuary	*Total habitat ha*	*Value 2020 R/ha*	*Present ecological state*
Bot/Kleinmond	1546.5	663105	3
Breë (Breede)	1789	1047850	2.5
Bushman's	340.9	4501143	2.5
Kariega	216	1639987	3
Keurbooms	674.7	2451241	1.5
Knysna	2286.7	2284345	2.5
Kowie	235.4	5538842	3
Sundays	483.3	1206374	3
Swartkops	926.7	534102	4

Source: Turpie and Clark (2007) for values, Van Niekerk et al. (2019) for present ecological state. Turpie and Clark's (2007) values converted to 2020 values using the Consumer Price Index. The ecosystem service values included: 1) Subsistence value, 2) Property value, 3) Tourism value and 4) Nursery value.

3.4.3 *Value of ecosystem improvements*

The value of ecosystem benefits from estuaries increased with decreases in the present ecological state (Figure 3.5).

Table 3.4 indicates that the net present value is highest for Sc. 3 (nutrient reduction and flow reduction), followed by Sc. 2 (flow reduction only). However, Sc. 1 (nutrient reduction only) also has a positive return on investment. Ultimately, the choice of scenario will be determined by what is feasible at the Swartkops based on a political process.

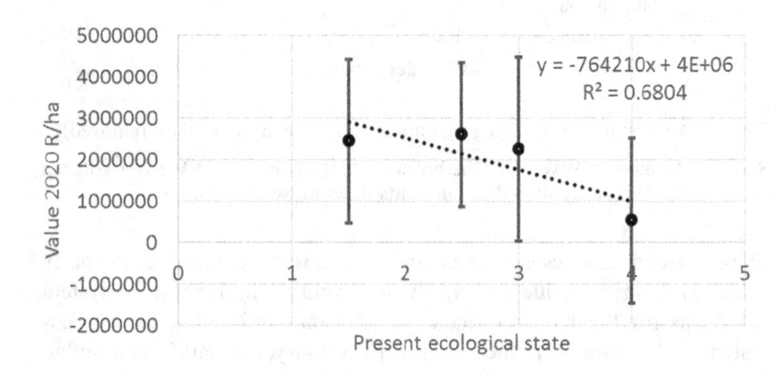

Figure 3.5 economic value of estuaries (2020 R/ha) based on the present ecological state.

Notes: error bars are standard deviations.

Table 3.4 Net present value of economic benefits (2020 billion rand) and benefit-cost ratios

	Baseline	*Sc1*	*Sc2*	*Sc3*
NPV @ 6% discount rate	0	1.2	4.798	6.996
Benefit-cost ratio (BCR)	0	4.0	n/a	35
Annualized value of benefits (2020 US$/ha/year)	0	5,957	23,817	34,728

Notes: See Table 3.1 for description of scenarios. Project time frame 25 years; Sc2 & Sc4 assumes that phytoremediation costs are 5% of conventional (WWTW) costs.

The baseline model assumes a discount rate of 6% and capital costs per tonne DIN of R1.28 million, but capital costs could be as high as R4.8 million per tonne DIN (Mitchell et al. 2014). Annual operating costs also are assumed to be 1.0% of capital costs under the baseline, but could be as high as 3.7% (Mitchell et al. 2014). We therefore conduct sensitivity analysis on the discount rate (range 4%–8%) as well as the capital cost of WWTWs (R1.28–4.8m per tonne DIN) as well as the annual operating costs (1%–3.7% of capital cost) using Monte Carlo simulation. The sample was drawn from 200 simulations using the random uniform distribution. The results, shown in Figure 3.6, show that only the flow reduction scenarios unilaterally demonstrate that the project is economically viable across all specifications. However, Sc.3 (nutrient reduction and flow reduction) provides the highest returns on investment.

3.5 Limitations of the study

It is important to note that the ecosystem values reported in Turpie and Clark (2007) are values for the top 20 most valuable estuaries for each ecosystem service (subsistence, property, tourism and nursery). The value estimates reported on may therefore not be indicative of values for South Africa in general, and may even overstate those values. It is therefore dangerous to attempt to extrapolate these values for other estuaries, and even to make inferences about the value of estuaries in South Africa from the work in this chapter. At the same time, these four values, although substantial, do not represent all values associated with estuaries in South Africa. For example, values such as water purification, protection against stormwater surges, existence values and carbon values are not included in the above assessment (although some of these values were valued by Turpie and Clark 2007), and therefore estuary values in general could be much higher than this.

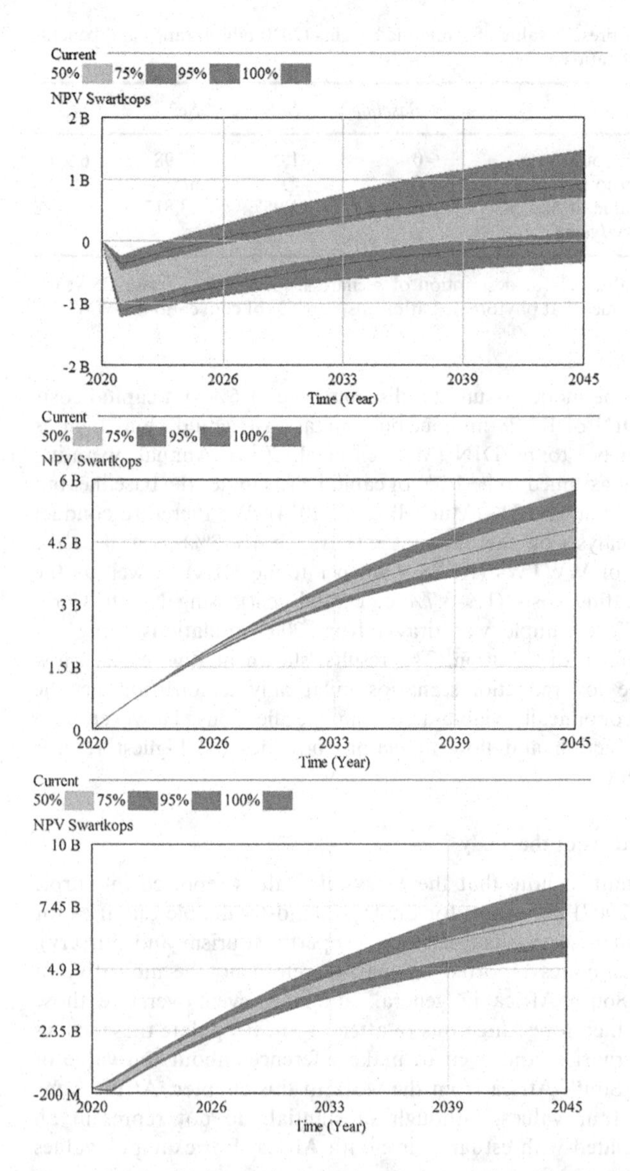

Figure 3.6 Sensitivity analysis of benefit-cost ratios (discount rate: 4%–8%; Capital costs (R1.28–4.8m/tonne DIN); Annual operating costs (1%–3.7% of capital costs). A. Sc1. Top; B. Sc2. Middle; C. Sc3. Bottom. For scenario descriptions see Table 3.1.

3.6 Conclusions

Swartkops estuary is a heavily degraded system, and solutions recommended by the National Biodiversity Assessment include improving river water quality, rehabilitating riparian areas/wetlands, and removing alien vegetation. In addition to the riparian area rehabilitation and removal of alien vegetation, two different options could be considered to improve river quality as it relates to nutrient loads: 1) capital investment to improve the three WWTWs; 2) flow reduction. The chapter develops a cost-benefit analysis using system dynamics modelling to assess these two options. Restoration costs are compared with the value of ecosystem improvements (improved property values, tourism values, subsistence values and nursery values).

The results of the cost-benefit analysis all indicate that both nutrient reduction and flow reduction policies are economically viable. Furthermore, while the scenarios with flow reduction produce higher returns, the optimal intervention is both a flow reduction strategy as well as a nutrient reduction strategy. Whether or not the scenario is politically acceptable remains to be seen.

References

Adams, J.B., Whitfield, A.K. and Van Niekerk, L. 2020. "A socio-ecological systems approach towards future research for the restoration, conservation and management of southern African estuaries." *African Journal of Aquatic Science* 45(1-2): 231–241.

Adams, J.B., Pretorius, L. and Snow, G.C. 2019. "Deterioration in the water quality of an urbanised estuary with recommendations for improvement." *Water SA* 45(1): 86–96.

Adams, J. and Riddin T. 2020. *State of knowledge: conservation and management of the Swartkops Estuary*. Port Elizabeth, South Africa: Institute for Coastal and Marine Research, Nelson Mandela University.

Adams J.B., Taljaard, S., Van Niekerk, L. and Lemley, D.A. 2020. "Nutrient enrichment as a threat to the ecological resilience and health of microtidal estuaries." *African Journal of Aquatic Science* 45: 23–40.

Andersson, E., Langemeyer, J., Borgström, S., McPhearson, T., Haase, D., Kronenberg, J., … and Baró, F. 2019. "Enabling green and blue infrastructure to improve contributions to human well-being and equity in urban systems." *BioScience* 69(7): 566–574.

Basconi, L., Cadier, C. and Guerrero-Limón, G. (2020). "Challenges in Marine Restoration Ecology: How Techniques, Assessment Metrics, and Ecosystem Valuation Can Lead to Improved Restoration Success." In Jungblut, S., Liebich, V. and Bode-Dalby, M. (Eds.). *YOUMARES 9-The Oceans: Our research, our future* (pp. 83–99). Cham: Springer.

Bester, R., Blignaut, J.N. and Crookes, D.J. 2019. "The impact of human behaviour and restoration on the economic lifespan of the proposed Ntabelanga and Laleni dams, South Africa: A system dynamics approach." *Water Resources and Economics* 26: 100126.

Brondizio, E.S., Settele, J., Díaz, S. and Ngo, H.T. 2019. *Global assessment report on biodiversity and ecosystem services of the Intergovernmental Science-Policy Platform on Biodiversity and Ecosystem Services*. Bonn, Germany: IPBES Secretariat.

Claassens L, Adams JB, Wasserman J, Whitfield AK. Under review. Estuary Restoration – A South African Perspective. Waltham N (Ed.). In: Repairing the World's Estuaries: Knowledge, Challenges and Possible Way Forward. Elsevier, Amsterdam.

Crookes, D.J., Blignaut, J.N., de Wit, M.P., Esler, K.J., Le Maitre, D.C., Milton, S.J., Mitchell, S.A., Cloete, J., de Abreu, P., Fourie (nee Vlok), H., Gull, K., Marx, D., Mugido, W., Ndhlovu, T., Nowell, M., Pauw, M., and Rebelo, A. 2013. "System dynamic modelling to assess economic viability and risk trade-offs for ecological restoration in South Africa." *Journal of Environmental Management* 120: 138–147.

Crookes, D.J., Blignaut, J.N. and Le Maitre, D.C. 2020. "The effect of accessibility and value addition on the costs of controlling invasive alien plants in South Africa: A three species system dynamics model in the fynbos and grassland biomes." *Southern Forests* 82(2): 125–134. 10.2989/20702620.2019.1686685

Department of Water Affairs (DWA) 2011. *Planning Level Review of Water Quality in South Africa; Sub-series no*. WQP 2.0; Pretoria, South Africa: DWA.

Department of Water Affairs (DWA) 2014. *2014 Green Drop Report*. Pretoria: Department of Water Affairs, South Africa.

Enviro-Fish Africa 2009. *C.A.P.E. Estuary Management Programme; Swartkops Integrated Environmental Management Plan: Draft Situation Assessment. Volume 1: Report to the Nelson Mandela Bay Municipality*. Grahamstown: Enviro-Fish Africa.

Evans, W. 2020. "What does water pollution cost? Towards a holistic understanding." *The Water Wheel* September/October: 42–45.

Freeman, L.A., Corbett, D.R., Fitzgerald, A.M., Lemley, D.A., Quigg, A. and Steppe, C.N. 2019. "Impacts of urbanization on estuarine ecosystems and water quality." *Estuaries and Coasts* 42: 1821–1838.

Graham, M., Blignaut, J., de Villiers, L., Mostert, D., Sibande, X., Gebremedhin, S., Harding, W., Rossouw, N., Freese, N.S., Ferrer, S. and Browne, M. 2012. *Development of a generic model to assess the costs associated with eutrophication*. Water Report Research Commission. Pretoria: South Africa.

Higgins, S.I., Turpie, J.K., Costanza, R., Cowling, R.M., Le Maitre, D.C., Marais, C. and Midgley, G.F. 1997. "An ecological economic simulation model of mountain fynbos ecosystems: dynamics, valuation and management." *Ecological Economics* 22(2): 155–169.

Lemley, D.A., Adams, J.B. and Strydom, N.A. 2017. "Testing the efficacy of an estuarine eutrophic condition index: Does it account for shifts in flow conditions?" *Ecological Indicators* 74: 357–370.

Lemley, D.A., Adams, J.B., Bornman, T.G., Campbell, E.E. and Deyzel, S.H.P. 2019. "Land-derived inorganic nutrient loading to coastal waters and the potential implications for nearshore plankton dynamics." *Continental Shelf Research* 174: 1–11.

Lemley, D., 2020. Personal communication. Email correspondence.

Malone, T.C. and Newton, A. 2020. “The globalization of cultural eutrophication in the coastal ocean: Causes and consequences.” *Frontiers in Marine Science* 7: 670.

Mitchell, S.A., De Wit, M.P., Blignaut, J.N. and Crookes, D. 2014. “Waste water treatment plants: The financing mechanisms associated with achieving green drop rating.” Water Research Commission Report Number WRC Report No. 2085/1/14. Pretoria.

Mudavanhu, S., Blignaut, J.N., Vink, N., Crookes, D., Meincken, M., Effah, B., Murima, D. and Nkambule, N. 2017a. “An assessment of the costs and benefits of using *Acacia saligna* (Port Jackson) and recycled thermoplastics for the production of wood polymer composites in the Western Cape Province, South Africa.” *African Journal of Agricultural and Resource Economics* 12(4): 322–365

Mudavanhu, S., Blignaut, J.N., Vink N., Crookes, D. and Nkambule, N. 2017b. “An economic analysis of different land-use options to assist in the control of the invasive *Prosopis* (Mesquite) tree.” *African Journal of Agricultural and Resource Economics* 12(4): 366–411

NBA 2019. *South African National Biodiversity Assessment 2018: Technical Report. Volume 3: Estuarine Realm*. Pretoria: CSIR & SANBI.

Nelson Mandela University (NMU) 2020. *Water Research Commission Workshop: Innovative Approaches For Estuary Water Quality Improvement*. WRC Project C2020/2021-00076. 11 November 2020. Nelson Mandela Bay.

Nkambule, N.P., Blignaut, J.N., Vundla, T., Morokong, T. and Mudavanhu, S. 2017. “The benefits and costs of clearing invasive alien plants in northern Zululand, South Africa.” *Ecosystem Services* 27: 203–223.

Oberholster, P. 2020. “Low cost green technology for domestic wastewater treatment for reuse and beneficiation.” Presentation given at the WRC Workshop Innovative Approaches for Estuary Water Quality Improvement [online], 11 November 2020.

Olisah, C., Adeniji, A.O., Okoh, O.O. and Okoh, A.I. 2019. “Occurrence and risk evaluation of organochlorine contaminants in surface water along the course of Swartkops and Sundays River Estuaries, Eastern Cape Province, South Africa.” *Environmental Geochemistry and Health* 41(6): 2777–2801.

Olisah, C., Okoh, O.O. and Okoh, A.I. 2020. “Spatial, seasonal and ecological risk assessment of organohalogenated contaminants in sediments of Swartkops and Sundays Estuaries, Eastern Cape province, South Africa.” *Journal of Soils and Sediments* 20(2): 1046–1059.

Reyes-García, V., Fernández-Llamazares, Á., McElwee, P., Molnár, Z., Öllerer, K., Wilson, S.J. and Brondizio, E.S. 2019. “The contributions of Indigenous Peoples and local communities to ecological restoration.” *Restoration Ecology* 27(1): 3–8.

Sachs, J.D., Schmidt-Traub, G., Mazzucato, M., Messner, D., Nakicenovic, N. and Rockström, J. 2019. “Six transformations to achieve the sustainable development goals.” *Nature Sustainability* 2(9): 805–814.

Snow, G., Adams, J. and Snow, B. 2019. Swartkops Estuary Research Symposium – Improving estuary health for the delivery of multiple ecosystem services. SANCOR Newsletter Issue 224.

Taljaard, S., Slinger, J.H. and Van Niekerk, L. 2017. “A screening model for assessing water quality in small, dynamic estuaries.” *Ocean & Coastal Management* 146: 1–14.

Turpie, J.K., Forsythe, K.J., Knowles, A., Blignaut, J. and Letley, G. 2017. "Mapping and valuation of South Africa's ecosystem services: A local perspective." *Ecosystem Services* 27: 179–192.
Turpie, J. and Clark, B. 2007. *Development of a conservation plan for temperate South African estuaries on the basis of biodiversity importance, ecosystem health and economic costs and benefits.* C.A.P.E. Regional Estuarine Management Programme. Final Report; August 2007. 125pp.
van Aswegen, J.D., Nel, L., Strydom, N.A., Minnaar, K., Kylin, H. and Bouwman, H. 2019. "Comparing the metallic elemental compositions of Kelp Gull Larus dominicanus eggs and eggshells from the Swartkops Estuary, Port Elizabeth, South Africa." *Chemosphere* 221: 533–542.
Van Niekerk, L., Adams, J.B., Lamberth, S.J., MacKay, C.F., Taljaard, S., Turpie, J.K., Weerts S.P. and Raimondo, D.C. 2019 (eds). *South African National Biodiversity Assessment 2018: Technical Report. Volume 3: Estuarine Realm.* CSIR report number CSIR/SPLA/EM/EXP/2019/0062/A. Pretoria: South African National Biodiversity Institute. Report Number: SANBI/NAT/NBA2018/2019/Vol3/A.
Van Zyl, H. and De Wit, M.P. 2013. *Environmental impact assessment (EIA) for the proposed N3: Keeversfontein to Warden (De Beers Pass Section).* DEA ref. no. 12/12/20/1992. Environmental Resource Economics DRAFT Specialist Report.
Vundla, T., Blignaut, J.N. and Crookes, D. 2017. "Aquatic weeds: To control or not to control. The case of the Midmar Dam, KwaZulu-Natal, South Africa." *African Journal of Agricultural and Resource Economics* 12(4): 412–429
Young, T.P. 2000. "Restoration ecology and conservation biology." *Biological Conservation* 92(1): 73–83.
Zweig, C.L., Newman, S. and Saunders, C.J. 2020. "Applied use of alternate stable state modeling in restoration ecology." *Ecological Applications* 30(8): e02195.

Appendix 1. List of constants, values and references used in the SER model

Constant	*Value*	*Unit*	*Reference/comment*
A	−764210	Rand/hectare/Year	Regression coefficient. See Figure 3.5
area of estuary	926.7	hectare	NBA (2019)
Baseline EC	3.75	Dimensionless	NBA (2019)
SC2 EC	3.25	Dimensionless	Based on data provided by Daniel Lemley (2020)
SC3 EC	3	Dimensionless	Based on data provided by Daniel Lemley (2020)
SC1 EC	3.583	Dimensionless	Based on data provided by Daniel Lemley (2020)

(*Continued*)

Constant	*Value*	*Unit*	*Reference/comment*
discount rate	0.06	Dimensionless	Own calculation based on Van Zyl and De Wit (2013)
Capital cost WWTW per unit DIN	1.28e + 006	Rand/tonne	Own calculation. See also Figure 1.4
flow volume proportion	0.5	Dmnl	1 (baseline & Sc1)= initial flow characteristics; Sc2: 0.5 = 50% reduction in flow; Sc3: 0.5 = 50% reduction in flow
MWC daily flow volume	6600	m3/day	NMU ongoing monitoring
MWC DIN concentration	4.06/1e+006	tonne/m3	NMU ongoing monitoring
Number of days in a year	365	day/Year	Known
Operating cost WWTW per unit DIN	12500	Rand/tonne	Calculation based on Graham et al. (2012)
proportion DIN reduction	0.5	Dmnl	0 (baseline & Sc2) = no reduction in DIN; Sc1: 0.5 = 50% reduction in DIN; Sc3: 0.5 = 50% reduction in DIN
WWTW1 daily flow volume	21120	m3/day	2014 Green drop report DWA (2014)
WWTW1 DIN concentration	12.5/1e+006	tonne/m3	DWS monitoring data DWA (2011)
WWTW2 daily flow volume	4410	m3/day	2014 Green drop report DWA (2014)
WWTW2 DIN concentration	11/1e+006	tonne/m3	DWS monitoring data DWA (2011)
WWTW3 daily flow volume	1980	m3/day	2014 Green drop report DWA (2014)
WWTW3 DIN concentration	17.4/1e+006	tonne/m3	DWS monitoring data DWA (2011)
FINAL TIME	2045	Year	The final time for the simulation.
INITIAL TIME	2020	Year	The initial time for the simulation.

4 Economics of water hyacinth removal in the Swartkops estuary

4.1 Introduction

Water hyacinth (*Eichhornia crassipes*) is a major invader that has infested many waterways around the world (Téllez et al. 2008). It has also invaded the Swartkops estuary. Its distribution is dynamic and is dependent on river flows (flows greater than 5m.s-1 wash it away) as well as whether or not the estuary is open or closed. Data in Nunes et al. (2020) for the uTongaki Estuary show how area covered by hyacinth versus indigenous plant cover changes depending on these characteristics.

Also, increased nutrification (N, P) from wastewater treatment works (WWTWs) increases the spread of these plants (Nunes et al. 2020). Water hyacinth can displace the local seagrass (*Zostera capensis*) in Swartkops (Adams et al. 2020), but also impedes water transport and recreational activities, and can cause drownings of livestock and humans (Dodd 2020). It also increases water-related diseases such as bilharzia and malaria as it provides an ideal breeding ground for snails and mosquitos (Dodd 2020).

Dense mats of water hyacinth can decrease light penetration in the water column, leading to declines in dissolved water oxygen, which leads to decreases in phytoplankton communities (Dodd 2020). Plants themselves can lead to further deterioration of water quality, and decreased water flow (Dodd 2020).

Biocontrol can be effective (leaf boring beetles). Wilson et al. (2001) developed a predator prey model for water hyacinth and a weevil biocontrol agent. Also, they reported on carrying capacity and growth rates.

The Department of Environment, Forestry and Fisheries (DEFF) has spent R2bn on control of Invasive Alien Aquatic Plant Species (IAAPS), of which R42m has been on the chemical control of water hyacinth (Dodd 2020). The other four of the Big Bad Five IAAPS (water lettuce, kariba weed, red water fern and Parrot's feather) can be controlled purely through biocontrol (Dodd 2020).

DOI: 10.4324/9781032651675-4

The aim of this chapter is to provide a model to assess the economic benefits of clearing water hyacinth. The economic benefits of seagrass recovery are compared with the costs of water hyacinth removal, in a benefit-cost model.

4.2 Literature review

Water hyacinth has received a mixed review in the literature. Some studies have highlighted the benefits of water hyacinth for the removal of harmful nutrients (e.g. Jafari 2010). Others, on the other hand, have highlighted its problems. For example, Coetzee and Hill (2012) present research that shows that higher nutrient loads in the water body, for example from effluent in WWTWs, actually facilitate the spread of water hyacinth.

A number of studies have quantified the economic impact (cost) of water hyacinth. Water hyacinth forms dense mats, thereby reducing the ability to navigate water bodies and impeding fishing, boating, access to smallholder farms and affecting other water-based recreational activities (Kateregga and Sterner 2009; Honlah et al. 2019). The economic impact of the water hyacinth on transportation and fisheries in Lake Victoria was estimated to be approximately US $350 million per annum (Mkumbo and Marshall 2014).

Water hyacinth also increases evapotranspiration of water compared to where it is absent. Arp et al. (2017) conducted an economic evaluation of the value of water saved by the removal of water hyacinth in the Vaalharts irrigation scheme. This equates to an annual benefit of between R54 million and R1.18 billion, which is a substantial improvement.

Water hyacinth also displaces rare submerged macrophyte species and reduces benthic macroinvertebrate diversity (Coetzee et al. 2014). Water hyacinth therefore reduces phytoplankton in water bodies, potentially decreasing fish production of those species dependent on phytoplankton (Villamagna and Murphy 2010). There is evidence to suggest that water hyacinth has displaced the native and important seagrass (*Z. capensis*) in some parts of the Swartkops estuary (van Niekerk et al. 2019).

The spread of water hyacinth has also disrupted hydroelectric power generation in Lake Victoria (Kateregga and Sterner 2007) in Nalubaale dam in Uganda (Kulyanyingi 2002), and in Pakistan (Fawad et al. 2015).

A challenge for its management is its rapid growth (Wolverton and McDonald 1979). Studies in Ethiopia suggest that it can double its mass every five days (Degaga 2019). This makes it one of the fastest growing macrophytes in the world (Tobias et al. 2019).

Various studies (Jones and Cilliers 1999; van Wyk and van Wilgen 2002; Yigermal and Assefa 2019) have advocated against utilizing a mechanical clearing option only for the control of water hyacinth, but rather utilizing an integrated control approach that incorporates mechanical, chemical and biological mehods. This is because manual clearing (Enyew et al. 2020) and biocontrol alone (Hill and Olckers 2001; Hill and Coetzee 2017) have not been as effective. The integrated approach seems to be the most effective for the control of the weed (Hill and Coetzee 2008). Costs of control are therefore expected to be higher than would be the case for biocontrol only (van Wyk and van Wilgen 2002). Other studies have reported costs for mechanical, chemical and manual control and documented their relative effectiveness (Alimi and Akinyemiju 1990).

Owing to the challenges in eradicating water hyacinth, some studies have advocated sustainable utilization of water hyacinth by transforming the weed into various economic uses (Güereña et al. 2015). Examples include converting water hyacinth biomass into organic fertilizers or biomass-to-energy plants. Other studies (May et al. 2022) have advocated limiting its spread to new areas.

In contrast with *Spartina alterniflora*, another invasive aquatic plant (see Chapter 5), no studies were found that estimated the optimal control path for water hyacinth. This is important as it denotes the resources needed to control the weed. Given the potential economic benefits of the weed in terms of reduced nutrient loads in water, is it economically beneficial to control the weed? This is the topic of the present chapter.

4.3 Method

A model is developed for the Swartkops estuary using the system dynamics modelling approach alluded to earlier. In the model, changes in nutrient loads affect the spread of water hyacinth. There is no explicit population model, but estimates of rates of change in water hyacinth cover are influenced by whether or not the estuary is open, and nutrient concentrations.

Changes in nutrient loads from WWTWs are modelled using regression relationships from Nunes et al. (2020, their table 4), supplemented by water quality data for Swartkops estuary from DWA (2020). The regression relationship relates to flow, nutrient loads and whether or not the estuary is open or closed to the proportion of the area covered by the hyacinth (dependent variable).

The model includes the cost of integrated clearing, including the costs of clearing the catchment, which will increase the river flow and may help to keep the hyacinth away. The benefit is the value of the seagrass recovered (since once the water hyacinth is removed, it no longer impedes the growth of the seagrass).

The stock flow diagram for the system is given in Figure 4.1. A variable that is unknown in the proportion of water hyacinth that is

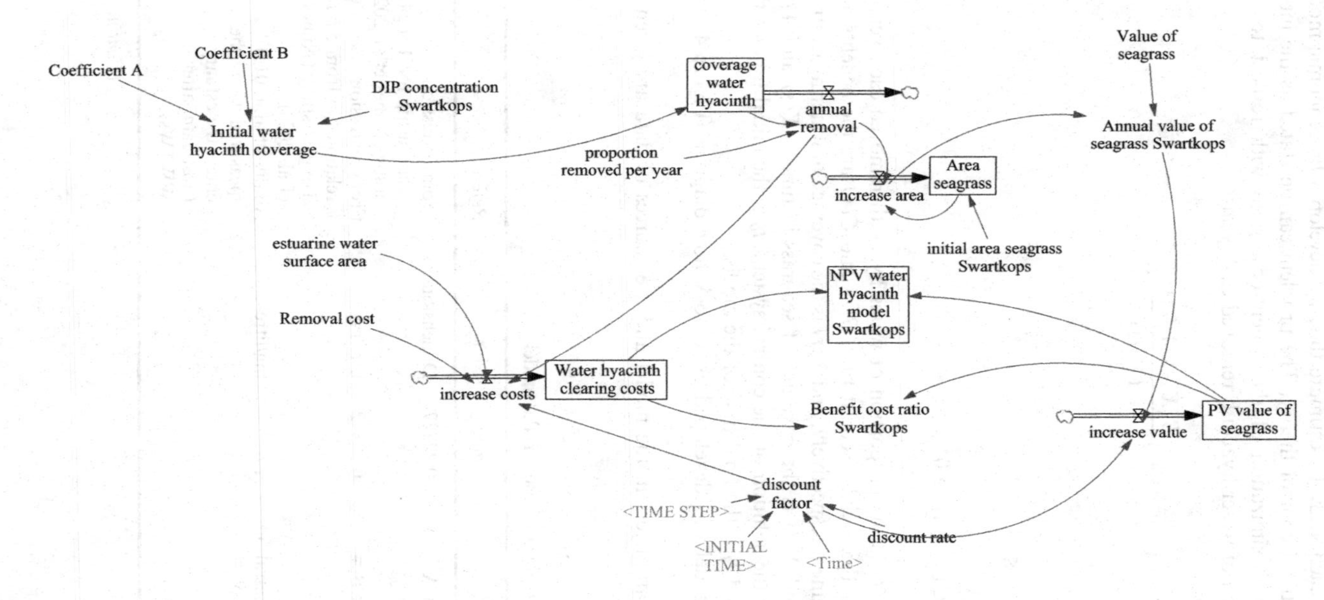

Figure 4.1 Stock flow diagram for the water hyacinth model.

removed each year. To estimate this, we develop a bioeconomic model using optimal control theory. The problem can be stated as the intertemporal maximization of the benefit-cost ratio with respect to the proportion of water hyacinth removed each year:

$$\max V(\cdot) \int_{t0}^{t1} e^{-\delta t} V(t) \frac{B(X(t), Y(t))}{C(X(t))} dt$$

Subject to

$$0 \le V(t) \le 1$$

$$B(t), C(t), X(t) \ge 0$$

Where $V(t)$ is the proportion of water hyacinth removed each year, δ represents the social discount rate, $B(\bullet)$ and $C(\bullet)$ are the benefits and costs in time t, respectively, and $X(t)$ is the coverage of water hyacinth in time t, and $Y(t)$ is the coverage of seagrass in time t. $X(\bullet)$ and $Y(\bullet)$ represent the solution of the control system, t_0 is the initial time of the model, and t_1 is the final time of the model.

The model was developed as a stock flow diagram (Figure 4.1) in Vensim.

The parameters in the model and the sources of data are given in Table 4.1.

Table 4.1 Constants used in the model

Constant	*Value*	*Unit*	*Reference*
Coefficient A =	0.24227	Dimensionless	From regression relationship for Tongaat study (Nunes et al. 2020)
Coefficient B =	0.23442	litre/mg	From regression relationship from data in Tongaat study (Nunes et al. 2020)
DIP concentration Swartkops =	1.8	mg/litre	Adams et al. (2019) measured where the river enters the estuary (upstream, after WWTWs)

(*Continued*)

Table 4.1 (Continued)

Constant	*Value*	*Unit*	*Reference*
Removal cost =	16178	Rand/hectare	2020 Rands (data from Crookes 2012) restoration costs (Kromme) – also D condition (like Swartkops)
estuarine water surface area =	135	hectare	Van Niekerk et al. (2019)
initial area seagrass Swartkops =	44.7	hectare	Bornman et al. (2016)
Value of seagrass =	198989	Rand/hectare	Lower bound based on data on value of seagrasses in Obura et al. (2017) converted to 2020 Rands
discount rate =	0.06	Dimensionless	Van Zyl and De Wit (2013)|
FINAL TIME =	2040	Year	The final time for the simulation
INITIAL TIME =	2020	Year	The initial time for the simulation.
TIME STEP =	1	Year [0,?]	The time step for the simulation.

4.4 Results

The value of V that maximizes the benefit-cost ratio is given in Table 4.2. Roughly a quarter of the water hyacinth should be removed each year to maximize societal benefits.

The optimal strategy is to reach the steady state X^* and Y^* following a "bang" (or most rapid) solution (Figure 4.2), but this trajectory is also influenced by the other variables in the model. The right graph shows a rapid decline in water hyacinth $(X(t))$ and the right graph shows a rapid increase in seagrass ($Y(t)$). The optimal steady state for water hyacinth (X^*) is zero and for seagrass (Y^*) is around 85 hectares for the Swartkops estuary. Removal of the water hyacinth will therefore

Table 4.2 Value of V that maximizes social benefit-cost ratio

Constant	*Value*	*Unit*	*Reference*
proportion removed per year (V) =	0.278053	1/Year	0–100% The value that maximizes the benefit-cost ratio was used

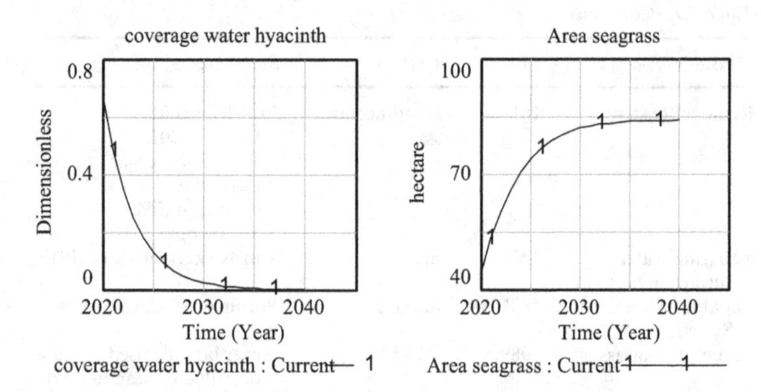

Figure 4.2 Optimal time path for $X(t)$ (left graph) and $Y(t)$ (right graph).

Table 4.3 NPV and benefit-cost ratio of optimal time path model

	Value/statistic
NPV @ 6% discount rate, over 20 years (2020 rands)	R 5.353 million
Benefit-cost ratio	5.235
Annualized value (2020 prices)	US$ 348/ha/yr

approximately double the area of seagrass in the Swartkops system, from 44.7 hectares to 85 hectares.

The optimal time path for the model produces a positive net present value and a benefit-cost ratio in excess of 1 (Table 4.3).

4.5 Model validation

Crookes (2022) showed that, for these types of models, structure verification, parameter verification, dimensional consistency and boundary adequacy tests are sine qua non. The model passed the dimensional consistency check (no unit errors), and passed the model verification test as well. The parameters are mostly derived from published sources, or estimated using the techniques described in this paper, and are deemed adequate. The model therefore has passed the main validation techniques and is suitable for use.

4.6 Discussion and conclusions

A novel feature of the present work is that the optimal time path for water hyacinth cleared is estimated using a bioeconomic model.

No other paper is known of that attempts this for water hyacinth, although there have been several papers to attempt this for the invasive marsh species *S. alterniflora* (see also Chapter 5). For example, Taylor and Hastings (2004) found that the optimal control strategy for *S. alterniflora* requires a large proportion of total area invaded to be cleared each year. This, as the authors note, requires a large amount of resources (labour, finances) to be committed to it. Here we find that a much smaller area of *E. crassipes* needs to be cleared on an annual basis for the species to be brought under control, requiring less labour and financial commitment. [Note, the differences in results between the present paper and the work of Taylor and Hastings (2004) is presumably due to the different biology of *S. alterniflora* and *E. crassipes*.]

The estimates of cost of rehabilitation used in this study (R16,178/ha in 2020 rands) is around double the costs for integrated clearing reported in van Wyk and van Wilgen (2002) of R807/ha (2020 values). The higher values used here are due to the fact that the costs of catchment rehabilitation are also included in the assessment. However, if the lower values of van Wyk and van Wilgen (2002) are used, the removal of water hyacinth becomes even more economically viable (the NPV would be higher).

Analyses such as these can help managers allocate scarce resources towards maximizing returns on restoration and IAAPS removal.

References

Adams, J.B., Pretorius, L. and Snow, G.C. 2019. "Deterioration in the water quality of an urbanised estuary with recommendations for improvement." *Water SA* 45(1): 86–96.

Adams, J.B., Taljaard, S., Van Niekerk, L. and Lemley, D.A. 2020. "Nutrient enrichment as a threat to the ecological resilience and health of South African microtidal estuaries." *African Journal of Aquatic Science* 45(1-2): 23–40.

Alimi, T. and Akinyemiju, O.A. 1990. "An economic analysis of water hyacinth control methods in Nigeria." *Journal of aquatic plant management* 28: 105–107.

Arp, R.S., Fraser, G.C.G. and Hill, M.P. 2017. Quantifying the economic water savings benefit of water hyacinth (Eichhornia crassipes) control in the Vaalharts Irrigation Scheme. *Water SA* 43(1): 58–66.

Bornman, T.G., Schmidt, J., Adams, J.B., Mfikili, A.N., Farre, R.E. and Smit, A.J. 2016. "Relative sea-level rise and the potential for subsidence of the Swartkops Estuary intertidal salt marshes, South Africa." *South African Journal of Botany* 107: 91–100. 10.1016/j.sajb.2016.05.003

Coetzee, J.A. and Hill, M.P. 2012. "The role of eutrophication in the biological control of water hyacinth, Eichhornia crassipes, in South Africa." *BioControl* 57, 247–261. 10.1007/s10526-011-9426-y

Coetzee J.A., Jones R.W. and Hill M.P. 2014. "Water hyacinth, Eichhornia crassipes (Pontederiaceae), reduces benthic macroinvertebrate diversity in a protected subtropical lake in South Africa." *Biodiversity and Conservation* 23: 1319–1330. 10.1007/s10531-014-0667-9

Crookes, D.J. 2012. "*Modelling the ecological-economic impacts of restoring natural capital, with a special focus on water and agriculture, at eight sites in South Africa.*" Doctoral dissertation, Stellenbosch University, Stellenbosch.

Crookes, D.J. 2022. *Mathematical models and environmental change: case studies in long term management*. Abingdon: Routledge. 10.4324/9781003247982

Degaga A.H. 2019. "Water hyacinth (Eichhornia crassipes) biology and its impacts on ecosystem, biodiversity, economy and human well-being." *Journal of Natural Sciences Research* 9(12): 24–30. doi:10.7176/JNSR

Dodd, C. 2020. *Invasive aquatic alien plants in South African estuaries* URL: https://youtu.be/QvvyZnhRnEI (Accessed: 18 January 2023)

DWA 2020. *Water quality monitoring data by water management area*: http://www.dwa.gov.za/iwqs/wms/data/WMS_WMA_txt.asp

Enyew, B.G., Assefa, W.W., and Gezie, A. 2020. "Socioeconomic effects of water hyacinth (Echhornia Crassipes) in Lake Tana, North Western Ethiopia." *PLoS ONE* 15(9): e0237668. 10.1371/journal.pone.0237668

Fawad, M., Khan, H., Gul, B., Khan, M.A. and Marwat, K.B. 2015. "Comparative effect of herbicidal and non-chemical control methods against water hyacinth." *Pakistan Journal of Weed Science Research* 21(4): 593–605.

Güereña D., Neufeldt H., Berazneva J., and Duby S. 2015. "Water hyacinth control in Lake Victoria: Transforming an ecological catastrophe into economic, social, and environmental benefits." *Sustainable Production and Consumption* 3: 59–69. 10.1016/j.spc.2015.06.003

Hill, M. and Coetzee, J. 2017. "The biological control of aquatic weeds in South Africa: Current status and future challenges." *Bothalia-African Biodiversity & Conservation* 47(2): 1–12.

Hill, M.P., and Coetzee J.A. 2008. "Integrated control of water hyacinth in Africa." *OEPP/EPPO Bulletin* 38: 452–457.

Hill M.P. and Olckers, T. 2001. "Biological control initiatives against water hyacinth in South Africa: constraining factors, success and new courses of action." In Julien, M.H., Hill, M.P., Center, T.D. and Ding, J. (Eds). Proceedings of the second Global Working Group meeting for the biological and integrated control of water hyacinth. Beijing, China, 9–12 October 2000, pp. 33–38. Australian Centre for International Agricultural Research, Canberra (AU).

Honlah, E., Segbefia, A.Y., Appiah, D.O. and Mensah, M. 2019. "The effects of water hyacinth invasion on smallholder farming along River Tano and Tano Lagoon, Ghana." *Cogent Food & Agriculture* 5(1): 1567042, DOI: 10.1080/23311932.2019.1567042

Jafari, N. 2010. "Ecological and socio-economic utilization of water hyacinth (Eichhornia crassipes Mart Solms)." *Journal of Applied Science and Environmental Management* 14(2): 43–49.

Jones, R. and Cilliers, C.J. 1999. "Integrated control of water hyacinth on the Nseleni /Mposa rivers and Lake Nsezi in KwaZulu-Natal, South Africa." In Hill, M.P., Julien, M.H. and Center, T.D. (Eds). Proceedings of the first IOBC Global Working Group meeting for the biological and integrated control of water hyacinth. 16–19 November 1998, Harare, Zimbabwe.

Kateregga, E. and Sterner, T. 2009. "Lake Victoria fish stocks and the effects of water hyacinth." *The Journal of Environment & Development* 18: 62–78.

Kateregga, E. and Sterner, T. 2007. "Indicators for an invasive species: Water hyacinths in Lake Victoria." *Ecological Indicators* 7: 362–370.

Kulyanyingi, V. 2002. *Economic losses/gains attributed to the water hyacinth.* Uganda: Fisheries Resources Research Institute (FIRRI) Jinja.

May, L., Dobel, A-J and Ongore, C. 2022. "Controlling water hyacinth (Eichhornia crassipes (Mart.) Solms): a proposed framework for preventative management." *Inland Waters* 12(1): 163–172, DOI: 10.1080/20442041.2021.1965444

Mkumbo, O. and Marshall, B. 2014. "The Nile perch fishery of Lake Victoria: Current status and management challenges." *Fisheries Management & Ecology* 22: 56–63.

Nunes, M., Adams, J.B. and Van Niekerk, L. 2020. "Changes in invasive alien aquatic plants in a small closed estuary." *South African Journal of Botany* 135: 317–329.

Obura, D., Gudka, M., Abdou Rabi, F., Bacha Gian, S., Bijoux, J., Freed, S., Maharavo, J., Mwaura, J., Porter, S., Sola, E. and Wickel, J. 2017. "Coral reef status report for the Western Indian Ocean." *Global Coral Reef Monitoring Network (GCRMN)/International Coral Reef Initiative (ICRI)*. pp 144.

Taylor, C.M. and Hastings, A. 2004. "Finding optimal control strategies for invasive species: A density-structured model for *Spartina alterniflora*." *Journal of Applied Ecology* 41: 1049–1057.

Téllez, T.R., de Rodrigo, E.M.L., Granado, G.L., Pérez, E.A., López, R.M. and Guzmán, J.M.S. 2008. "The water hyacinth, Eichhornia crassipes: An invasive plant in the Guadiana River Basin (Spain)." *Aquatic Invasions* 3: 42–53.

Tobias, V.D., Conrad, J.L., Mahardja, B., and Khanna, S. 2019. "Impacts of water hyacinth treatment on water quality in a tidal estuarine environment." *Biological Invasions* 21: 3479–3490.

Van Niekerk, L., Adams, J., Stephen, L., MacKay, F., Taljaard, S., Jane, T. and Weerts, S. 2019. *South African National Biodiversity Assessment 2018: Technical Report. Volume 3*. Estuarine Realm.

van Wyk, E. and van Wilgen, B.W. 2002. "The cost of water hyacinth control in South Africa: A case study of three options." *African Journal of Aquatic Science* 27: 141–149. 10.2989/16085914.2002.9626585

Van Zyl, H. and De Wit, M. 2013. "Environmental Impact Assessment for the proposed National Road 3: Keeversfontein To Warden (De Beers Pass Section) DEA Ref. No. 12/12/20/1992 Environmental Resource Economics Draft Specialist Report." Cape Town.

Villamagna, A.M. and Murphy, B.R. 2010. "Ecological and socio-economic impacts of invasive water hyacinth (Eichhornia crassipes): a review." *Freshwater Biol* 55(2): 282–298. 10.1111/j.1365-2427.2009.02294.x

Wilson, J.R., Rees, M., Holst, N., Thomas, M.B. and Hill, G. 2001. "Water hyacinth population dynamics. Biological and Integrated Control of Water Hyacinth, Eichhornia crassipes." ACIAR Proceedings No. 102.

Wolverton, B.C. and McDonald, R.C. 1979. "Water hyacinth (Eichhornia-crassipes) productivity and harvesting studies." *Economic Botany* 33: 1–10.

Yigermal, H. and Assefa, F. 2019. "Impact of the invasive water hyacinth (Eichhornia crassipes) on socio-economic attributes: A review." *Journal of Agriculture and Environmental Science* 4(2): 45–55.

5 The economic benefits of removing *Spartina alterniflora* from the Great Brak estuary

5.1 Introduction

Spartina alterniflora is an aggressive plant invader that has been declared a category 1 A invasive species in South Africa, implying that it should be eradicated where possible (Riddin et al. 2016). In the Great Brak estuary it has displaced intertidal salt marsh species such as *S. tegetaria, Triglochin spp., C. coronopifolia* and *B. diffusa*, as well as the intertidal seagrass *Zostera capensis Setchell* (Adams et al. 2012).

However, Wu and Wang (2011) indicate that at low densities, *S. alterniflora* improves soil conditions and is beneficial to native plant species. At high density, on the other hand, it competes with native plants for light and carbon dioxide. Over the longer term, the net effect of *S. alterniflora* on native plant species is negative.

Riddin et al. (2016) describe chemical treatment of *S. alterniflora*, and concluded that the plant had been "eradicated," but could still re-occur under certain conditions. One of those conditions that could influence its reoccurrence, is its high intrinsic growth rate. Du Toit and Campbell (2002), indicate that "*Spartina alterniflora Loisel* transplants of 1 m spacing along the east coast of North America took only two seasons to achieve complete cover" (p. 454).

Economic incentives are important for ensuring the restoration of natural capital (Larrosa et al. 2016; Milton et al. 2003). An absence of economic incentives inhibit restoration efforts. Therefore, the question is asked, are the economic conditions right for the eradication of the weed?

A competition mutualism model is developed, incorporating economic data and information on the spread of *S. alterniflora* in the Grak Brak estuary in order to attempt to replicate the conditions that brought the weed under control. The model is then projected forward in order to ascertain under what conditions the weed will persist or be eradicated.

DOI: 10.4324/9781032651675-5

5.2 Literature review

The economic benefits of planting *S. alterniflora* in the Tampa Bay area was discussed by Hoffman and Rodger (1980), but the weed is now recognized as a major problem species worldwide (Strong and Ayres 2009). Economic analysis for its removal is fairly widespread, but most seek to estimate the optimal removal rate. For example, Jardine and Sanchirico (2018) show that a "big bang" approach is optimal, where the majority of the plants are removed in the first year (see also Buhle et al. 2005), or a single age class is targeted (Hastings et al. 2006).

Other studies (Epanchin-Niell and Hastings, 2010; Grevstad 2005; Taylor and Hastings 2004) have shown that targeting low density stands early on in the invasion is optimal. This is the approach that was adopted in the Great Brak area (Riddin et al. 2016). A relatively new strand of Spartina was discovered in the Great Brak estuary in 2004, and treated with herbicide between 2013 and 2015, and by November 2015 was virtually eradicated (Riddin et al. 2016).

At the same time, little is known about the economic drivers for Spartina removal in the Great Brak estuary. Was this decision to clear based on economic considerations, or was the motivation purely political? This chapter seeks to explore these issues to a greater extent.

5.3 Methodology

A stock flow diagram for the system is constructed (Figure 5.1). The model is constructed in the system dynamics modelling package Vensim. A detailed exposition of the equations in the system is given in the next section.

5.3.1 Equations

Spartina interacts with native species in a system such that, at low densities there is mutualism characterizing the interactions, but at higher densities there is competition (Wang and Wu 2011). We utilize as the basis for our system the mutualism/competition model of Wang and Wu (2011), which we have modified by adding an additional interaction term:

$$\frac{dX_1}{dt} = r_1 X_1 \left(1 + \frac{a_1 b_2}{K_1} - \frac{X_1}{K_1}\right) - \frac{r_1 a_1}{K_1}|X_2 - b_2|X_1 + q_1 E_1 X_1$$

$$\frac{dX_2}{dt} = r_2 X_2 \left(1 + \frac{a_2 b_1}{K_2} - \frac{X_2}{K_2}\right) - \frac{r_2 a_2}{K_2}|X_1 - b_1|X_2 - q_2 E_2 X_2$$

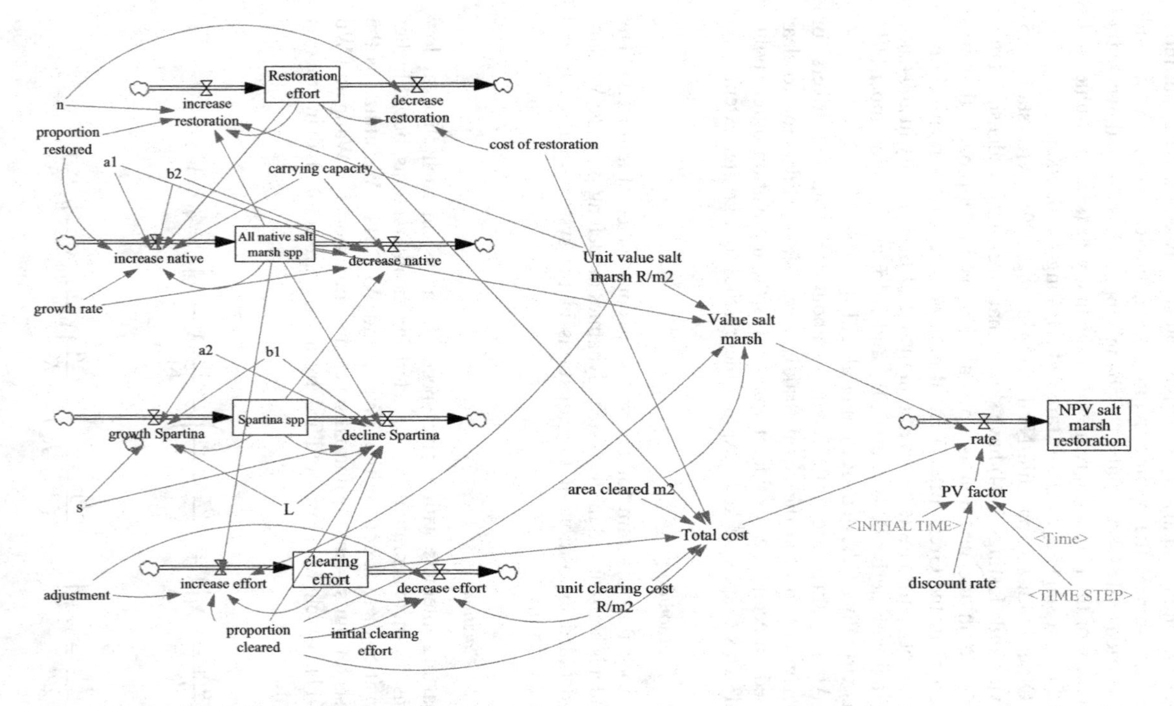

Figure 5.1 Stock flow diagram for the system.

Where X_1 is the native marsh species population density (ranging between 0 and 1), X_2 is the Spartina population density (ranging between 0 and 1), r_1 and r_2 are the intrinsic growth rates for native and Spartina species, respectively, and K_1 and K_2 are their carrying capacities. The other parameters in the model (a_1, a_2, b_1, b_2) are interaction terms that influence the extent to which X_1 impacts X_2, and vice versa.

The model has two further differential equations that impact the system. The first differential equation models a mutualistic relationship between restoration effort and native marsh species. In other words, the higher the net revenues from restoration, the more likely it is that marsh species will be restored. It is assumed that some form of active restoration is required to re-establish native plant species. The equation for the system is as follows:

$$\frac{dE_1}{dt} = n_1(pq_1E_1 - c_1E_1)$$

The parameter n_1 is an adjustment coefficient relating net revenues from restoration to restoration effort, p is the unit value of marsh species, q_1 is the proportion restored in a given year, and E_1 is the restoration effort.

The second differential equation is a Lotka-Volterra equation that affects how net returns from clearing effort expended on Spartina removal influences the removal of Spartina:

$$\frac{dE_2}{dt} = n_2(pq_2E_2 - c_2E_2)$$

Which is the standard fishery bioeconomics differential equation, but this time focused on invasive species removal effort. The parameter n_2 is an adjustment coefficient relating changes in profits to clearing effort, and p is the unit value of salt marsh species, q_2 is the proportion of Spartina cleared each year, E_2 is the clearing effort (ranging between 0 and 1) and c_2 is the unit clearing cost (the reader is referred to Table 5.2 for full definitions of these variables, their values in the model, and also the units used).

The final equation in the model is the Net Present Value (NPV) calculation:

$$\text{NPV} = \sum_{t=0}^{n} \frac{R_t - C_t}{(1 + i)^t}$$

where:

NPV = net present value

R_t, C_t = cash inflow (benefits) or outflows (costs) during a single period t

i = social discount rate

n = number of time periods

The cost component includes both the clearing costs (of Spartina) and the restoration costs (of native marsh species) (Table 5.1).

Table 5.1 Equations used in the model

Name	*Equation*	*Units*
Value salt marsh=	"Unit value salt marsh R/m2"*All native salt marsh spp*area cleared m2*proportion cleared	Rand/Year
increase effort=	"Unit value salt marsh R/ m2"*adjustment*All native salt marsh spp*proportion cleared*clearing effort	1/Year
increase native=	growth rate*All native salt marsh spp+ growth rate*All native salt marsh spp*a1*b2/carrying capacity+ proportion restored*Restoration effort*All native salt marsh spp	1/Year
increase restoration=	n*"Unit value salt marsh R/m2" *proportion restored*Restoration effort*All native salt marsh spp	1/Year
Restoration effort=	INTEG (increase restoration-decrease restoration, 0.001)	Dmnl
decrease restoration=	n*cost of restoration*Restoration effort	1/Year
All native salt marsh spp=	INTEG (increase native-decrease native,1)	Dmnl
Spartina spp=	INTEG (growth Spartina-decline Spartina,0.001)	Dmnl
growth Spartina=	s*Spartina spp+s*Spartina spp*a2*b1/L	1/Year
decrease native=	growth rate*All native salt marsh spp^2/ carrying capacity+growth rate*All native salt marsh spp/carrying capacity*a1*ABS(Spartina spp -b2)	1/Year
decline Spartina=	s*Spartina spp^2/L+s*Spartina spp*a2/ L*ABS(All native salt marsh spp-b1) +clearing effort*Spartina spp*proportion cleared	1/Year
decrease effort=	adjustment*"unit clearing cost R/m2" *clearing effort*proportion cleared	1/Year
rate=	(Value salt marsh-cost of clearing)*PV factor	Rand/Year
clearing effort=	INTEG (increase effort-decrease effort, initial clearing effort)	Dmnl
cost of clearing=	clearing effort*area cleared m2* "unit clearing cost R/m2"*proportion cleared	Rand/Year
PV factor=	(1/(1+discount rate))^((Time-INITIAL TIME)/TIME STEP)	Dmnl
NPV salt marsh restoration=	INTEG (rate,0)	Rand

Table 5.2 Constants used in the model, values, units and description

Parameter	*Value*	*Unit*	*Description (reference)*
proportion restored=	0.94	1/Year	Proportion of total area of native marsh spp restored in a given year. Model calibrated estimate to obtain best fit with historical data.
n=	0.01	m2/Rand	This is an adjustment coefficient that relates the change in profits in the year to proportion restored.
cost of restoration=	0.75	Rand/(m2*Year)	The cost to restore native marsh spp after removal of Spartina Spp. Model calibrated to obtain best fit with the historical data.
a1=	15	Dmnl	Parameters in the Wang and Wu competition/mutualism model (see text)
a2=	2.134	Dmnl	Parameters in the Wang and Wu competition/mutualism model (see text)
b1=	1	Dmnl	Parameters in the Wang and Wu competition/mutualism model (see text)
b2=	1	Dmnl	Parameters in the Wang and Wu competition/mutualism model (see text)
carrying capacity=	1	Dmnl	Carrying capacity for the native marsh spp model. A value of 1 indicates that the native spp could occupy the entire area.
growth rate=	−0.04295	1/Year	Intrinsic growth rate for the native marsh spp. Value of the parameter that gave the best fit with the historical data.
adjustment=	0.01	m2/Rand	This is an adjustment coefficient that relates the change in profits in the year to proportion cleared.
initial clearing effort=	1	Dmnl	Proportion of total area of Spartina spp. cleared in the first year
proportion cleared=	0.32	1/Year	Proportion of total area of *S. alterniflora* cleared in a given year. The model calibrated estimate to obtain best fit with historical data.
area cleared m2=	12620	m2	The maximum forecasted area *S. alterniflora* occupied in 2012, before clearing commenced in 2013 (Riddin et al. 2016)

(*Continued*)

Table 5.2 (Continued)

Parameter	*Value*	*Unit*	*Description (reference)*
discount rate=	0.06	Dmnl	The value of the social rate of time preference, used in net present value calculations (Calculated based on the method reported in Van Zyl and De Wit 2013)
"unit clearing cost R/m2"=	12.7	Rand/m2	Cost of removing *S. alterniflora* in Great Brak, converted to 2019 values (Riddin et al. 2016)
"Unit value salt marsh R/m2"=	76	Rand/m2	2019 values literature value = 11.5; value above is the minimum value needed to replicate the data and eradicate Spartina.
L=	1	Dmnl	Carrying capacity of logistic model. A value of 1 indicates that *S. alterniflora* could occupy the entire area (which indeed it did in 2012).
s=	0.7979	1/Year	Intrinsic growth rate of *S. alterniflora*. The model calibrated value that best fitted the historical data.
FINAL TIME =	2040	Year	The final time for the simulation
INITIAL TIME =	2002	Year	The initial time for the simulation
TIME STEP =	1	Year [0,?]	The time step for the simulation

5.3.2 *Constants*

A list of the constants used in the model, their units, a description and a data source are given in Table 5.2.

5.4 Results

There are three scenarios in the model:

Scenario A. Baseline. Value of seagrass as per literature (R11.5/m^2) and unit clearing cost R12.7/m^2. Under this scenario Spartina is not eradicated but ultimately converges on its carrying capacity. The native marsh species are reduced to zero, and net present value from clearing and restoration is negative (Figure 5.2, top row).

Scenario B. Value of seagrass increased to R76/m^2 and all other variables in the model are kept constant. This scenario results in better replication of the system: the native marsh species rebound to around 50% of carrying capacity, but the weed (Spartina) is still not eradicated and also recovers to around 50% of carrying capacity. Net present values are zero (Figure 5.2, middle row).

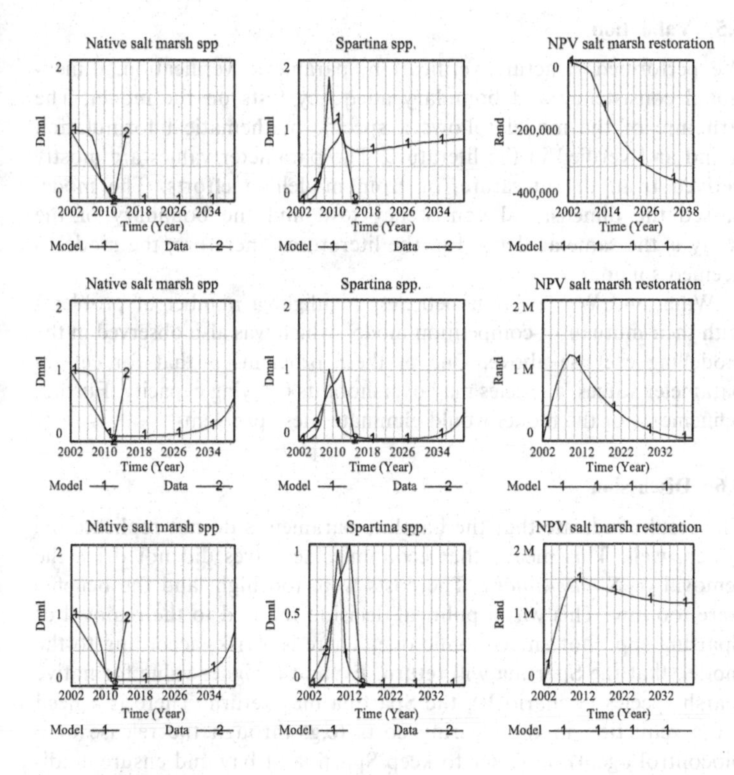

Figure 5.2 Top row. Baseline (scenario A). Middle row: scenario B. Bottom row: scenario C. Description of scenarios given in the text.

Scenario C. The value of native species increased to R70/m^2 and clearing costs reduced to R4/m^2 (e.g. through the introduction of bio-control agent). Under this scenario, Spartina is eradicated and native marsh species recover to around 70% of carrying capacity by 2040. The net present value is around R1.25 million (over 38 years, at a discount rate of 6%) (Figure 5.2, bottom row).

Results of the Net Present Value calculations for the different scenarios are summarized in Table 5.3.

Table 5.3 Economic benefits of salt marsh restoration

Scenario	*A*	*B*	*C*
NPV (2019 R million)	−0.4	0	1.25
Annualized values (US$/ha/year)	−1,334.35	0.00	4,169.85

NPVs are at a 6% % discount rate, over 38 years

5.5 Validation

We performed structure verification, parameter verification, dimensional consistency and boundary adequacy tests on the model. The structure of the model follows a specific mathematical formulation found and verified in the literature. The parameter values are mostly derived from the literature, or from modelling efforts. The model passed the dimensional consistency test, and the boundary of the study is the same as defined in the literature. Therefore, the model is deemed sound.

Wang and Wu (2011) nonetheless highlight a number of problems with their mutualism-competition model, which was also observed in the modelling exercise above. One of these problems is that for certain parameter values, a species may overshoot its carrying capacity. Further refinements of the model would eliminate these problems.

5.6 Discussion

The results indicate that the baseline parameters do not replicate the system well. This means that economic incentives did not drive the removal of *S. alterniflora*. The costs were too high, and the benefits were too low. This was a political decision that led to the removal of Spartina spp., not an economic one. There is a risk, according to the model, that the Spartina will return. Even at a higher value for native marsh species (scenario B), the Spartina may return. There is a need for a reduction in the clearing costs (e.g. through the release of a biocontrol agent), in order to keep Spartina at bay and ensure eradication of the species. Positive economic returns to restoration are possible, but will require an improvement in value of benefits and reduction in costs of removal.

Are the conditions that result in the controlling of *S. alterniflora* realistic? Estimates of the baseline value of salt marsh species (R11.5/m^2), which were derived based on data for salt marshes in Blignaut et al. (2017), seem to be fairly typical. Estimates of the value of salt marshes reported in Barbier et al. (2011), based on three annual ecosystem services valued – raw materials and food, coastal protection, and carbon sequestration – produced similar values (R12.08/m^2 in 2019 prices). However, they also reported on other values that, although not annualized, could increase the value of salt marshes. These include water purification; maintenance of fisheries (nursery values); and tourism, recreation, education and research. Estimating annual values from these capitalized values could increase the value of salt marshes by R4/m^2 (R3/m^2 for water purification, R1/m^2 for maintenance of fishes, and recreational benefits not estimated). So, it is unlikely that a sufficiently high value for salt marshes could be achieved to make clearing economically viable.

Table 5.4 Costs of biocontrol (2019 Rands/m^2) for selected invasive species in South Africa

Invasive species	*Cost of biocontrol 2019 R/m^2*
Jointed cactus	0.001
Red sesbania	0.004
Lantana	0.003
Long-leaved wattle	0.004
Golden wattle	0.020
Silky hakea	0.004
Water hyacinth	0.087

Source: First six species, estimated from van Wilgen et al (2004); Water hyacinth – van Wyk and van Wilgen (2002).

However, there are costs of biocontrol for *S. alterniflora* in South Africa, but comparisons with the biocontrol costs for other invasive plants (Table 5.4) indicate that these costs are potentially much less than those reported in this study, even negligible in some instances when expressed in per m^2.

Reducing the biocontrol cost to zero in the model means that the economic value of marsh species would need to be R67/m^2 to eradicate Spartina, but this is likely still too high a value for salt marsh species to make the economic option viable. In this case, the eradication of Spartina is likely based on a political decision rather than an economic one.

A further consideration is that the present model is simply insufficiently robust to model this system. Perhaps it attempts to be too complicated and a simpler model would suffice. Even so, there are some conditions where an eradication of Spartina and restoration of native marsh species are possible, and fit the historical data reasonably well.

5.7 Conclusions

This chapter models a mutualistic relationship between restoration effort and increase in salt marsh spp. The model does not predict a recovery in salt marsh species at baseline values (R11.5/m^2 price and R12.7/m^2 cost), in contrast to the literature, but if price increased and cost of clearing Spartina is reduced, then salt marsh species recover and Spartina is eliminated.

This study shows that the decision to clear, although based possibly on the international economic literature on optimal clearance rates, does not make sense from a local economic perspective. However, the economic criterion is not the only requirement for alien species removal. Political expediency is an important and often overlooked criterion.

References

Adams, J.B., Grobler, A., Rowe, C., Riddin, T., Bornman, T.G. and Ayres, D.R. 2012. "Plant traits and spread of the invasive salt marsh grass, Spartina alterniflora Loisel., in the Great Brak Estuary, South Africa." *African Journal of Marine Science 2012* 34(3): 313–322. 10.2989/1814232X.2012.725279

Barbier, E.B., Hacker, S.D., Kennedy, C., Koch, E.W., Stier, A.C. and Silliman, B.R. 2011. "The value of estuarine and coastal ecosystem services." *Ecological Monographs* 81: 169–193. 10.1890/10-1510.1.

Blignaut, J., Mander, M., Inglesi-Lotz, R., Glavan, J. and Parr, S. 2017. "Economic value of the Abu Dhabi coastal and marine ecosystem services: Estimate and management applications." In Azar, E. and Raouf, M.A. (Eds.), 2017. *Sustainability in the Gulf: Challenges and opportunities.* (pp. 210–227). Routledge.

Buhle, E.R., Margolis, M. and Ruesink, J.L. 2005. "Bang for buck: cost-effective control of invasive species with different life histories." *Ecological Economics* 52: 355–366.

De Wit, M.P. and Van Zyl, H. 2013. Environmental Impact Assessment for the proposed National road 3: Keeversfontein to Warden (De Beers pass section) DEA ref. no. 12/12/20/1992Environmental Resource Economics DRAFT Specialist Report. Prepared for Cave Klapwijk and Associates.

Du Toit, S.R. and Campbell, E.E. 2002. "An analysis of the performance of an artificial wetland for nutrient removal in solar saltworks." *South African Journal of Botany* 68: 451–456.

Epanchin-Niell, R.S. and Hastings, A. 2010. "Controlling established invaders: Integrating economics and spread dynamics to determine optimal management." *Ecology letters* 13(4): 528–541.

Grevstad, F.S. 2005. "Simulating control strategies for a spatially structured weed invasion: Spartina alterniflora (Loisel) in Pacific Coast estuaries." *Biological Invasions* 7: 665–677.

Hastings, A., Hall, R.J. and Taylor, C.M. 2006. "A simple approach to optimal control of invasive species." *Theoretical Population Biology* 70(4): 431–435.

Hoffman, W.E. and Rodger Jr, J.A. 1980. "A cost/benefit analysis of two large coastal plantings in Tampa Bay, Florida." In Wetlands Conference Proceeding. Tampa, Fla. EUA.

Jardine, S.L. and Sanchirico, J.N. 2018. "Estimating the cost of invasive species control." *Journal of Environmental Economics and Management* 87: 242–257.

Larrosa, C., Carrasco, L.R. and Milner-Gulland, E.J. 2016. "Unintended feedbacks: Challenges and opportunities for improving 1 conservation effectiveness." *Conservation Letters* 9(5): 316–326.

Milton, S.J., Dean, W.R.J. and Richardson, D.M., 2003. "Economic incentives for restoring natural capital in Southern African Rangelands." *Frontiers in Ecology and the Environment* 1(5): 247–254.

Riddin, T., Van Wyk, E. and Adams, J., 2016. "The rise and fall of an invasive estuarine grass." *South African Journal of Botany* 107: 74–79.

Strong, D.R. and Ayres, D.A. 2009. *Spartina introductions and consequences in salt marshes. Human impacts on salt marshes: a global perspective*. Berkeley and Los Angeles: University of California Press, pp.3–22.

Taylor, C.M. and Hastings, A. 2004. "Finding optimal control strategies for invasive species: a density-structured model for Spartina alterniflora." *Journal of Applied Ecology* 41: 1049–1057.

van Wilgen, B.W., de Wit, M.P., Anderson, H.J., Le Maitre, D.C., Kotze, I.M., Ndala, S. and Rapholo, M.B. 2004. "Costs and benefits of biological control of invasive alien plants: case studies from South Africa: working for water." *South African Journal of Science* 100(2) 113–122.

van Wyk, E. and van Wilgen, B.W. 2002. "The cost of water hyacinth control in South Africa: A case study of three options." *African Journal of Aquatic Science* 27(2): 141–149. 10.2989/16085914.2002.9626585

Wang, Y. and Wu, H. 2011. "A mutualism-competition model characterizing competitors with mutualism at low density." *Mathematical and Computer Modelling* 53: 1654–1663.

6 The economic benefits of restoring Knysna estuary following the 2017 Great Fire

6.1 Introduction

For six days, from 7 to 12 June 2017, fires blazed across the greater Knysna area. These fires have been likened to the Great Fire of 1869 (George Herald 2017). While there are piecemeal reports of the impact of the 2017 Knysna fire, there has to date not been a systematic assessment of the economic and financial costs at both a municipal and a national level.

A major reason for the spread of the fire was the prevalence of invasive alien plants. According to Stehle (2017), 180 km^2 of pine plantations were destroyed in the fire, or 25% of the total forestry area in the Eden district municipality. By contrast, less than 30 km^2 of indigenous forest was destroyed in the fire (Viljoen 2017). Geldenhuys (2017) argues that the reason that plantations and infrastructure were damaged in the recent fires was because property is built and exotic plant species are grown in fire prone areas, while the majority of indigenous forests occur in fire shadow areas. Studying the financial and economic impact of the Knysna fire provides a costing of the impact of such decisions and serves as a warning for future development. However, even without a costing, the need to eradicate invasive alien plants from fire prone areas has been identified (Mortlock 2017a). Furthermore, a costing study provides a means of assessing the benefits and costs accruing on different levels. For example, on a local level a fire such as this may actually be beneficial to the economy through job creation, stimulation of the local construction industry and the inflow of finances. This is likened to the economic benefits of removing or clearing invasive alien plants.

On a national level a fire like this results in a net societal loss. These losses decrease the ecosystem value of the estuary. It is important, therefore, to take into consideration both of these different levels of impacts.

In this chapter, the financial and economic impact of the 2017 Knysna fire is assessed. In the next section, the methodology is discussed, whereafter the results are presented. Finally, some conclusions and implications are drawn.

DOI: 10.4324/9781032651675-6

6.2 Methodology

Three levels of analysis are pursued, namely 1) a local financial analysis, 2) a national financial analysis, and 3) a national economic analysis. The data used for deriving the financial and economic flows are presented in Appendix A, Table A.1. The different categories of values used in the local financial analysis and national economic analysis are also provided (Table 6.1).

In this study, we distinguish between "above-the-line" and "below-the-line" impacts. Whether these are benefits or costs will depend on

Table 6.1 Categories used in the Knsyna municipality local financial and national economic analysis

Flow category	*Knysna local financial analysis*	*National economic analysis*
Private residences	x	x
Infrastructure [incl. roads, water, public buildings]	x	x
Other public sector reconstruction	x	x
Rates exemption fire victims	x	x
Total urban	x	
Agriculture (emergency measures, excl. feed)	x	x
Cost of fodder destroyed	x	x
Agriculture (loss of food production)	x	x
Timber (plantation, loss of production)	x	x
Emergency Response Operations [salaries only]	x	x
Environment – costs of restoration	x	x
Fatalities		x
Injuries		x
• Serious		x
• Minor [firefighters only]		x
Environment – loss of ecosystem services		x
• Fynbos		x
• Natural forest		x
Total costs		
Inflow		
Western Cape Provincial Government	x	x
Knysna Fire Disaster Relief Fund	x	x

(*Continued*)

Table 6.1 (Continued)

Flow category	*Knysna local financial analysis*	*National economic analysis*
Insurance claims [estimated]	x	
Other donations [not exhaustive]	x	x
National government contributions	x	x
Total inflow	x	
Construction multiplier	x	
Total inflow		
Net inflow (R mil.)		

whether the impact is at a local level or national level. At a local (i.e. Knysna) level, above-the-line impacts are costs, while below-the-line impacts are benefits. For national-level financial analysis, above-the-line impacts are benefits, while below-the-line impacts are costs. However, for national-level economic analysis, above-the-line impacts are also costs, which includes a number of other impacts such as the cost of fatalities and injuries, and loss of ecosystem goods and services.

In addition to this, the local (Knysna) financial costing includes a construction multiplier of 2.84 that captures the effect of infrastructure rebuilding on downstream industries in the construction and related industries. Construction impacts, timber and ecosystem loss are assumed to accrue over two years. The rest accrue over one year. This makes it possible to assess the impact of the fires on the Knysna Gross Domestic Product (GDP). In the next section, the results are presented.

6.3 Results

6.3.1 Local financial analysis

The local (Knysna) financial analysis indicates a net benefit to the local economy of between R1.8 and R2.5 billion (2017 prices) in 2017, also taking into consideration the construction multiplier (see Table 6.2). This equates to an injection into the Knysna GDP of between 38% and 53% in 2017. In 2018, benefits are lower but still substantial.

6.3.2 National financial analysis

On the national level, net inflows are still positive from a financial perspective. However, it is much lower than at the Knysna municipal level (see Table 6.3). This is largely due to the absence of a construction multiplier on a national level.

Table 6.2 Net financial inflows: Knysna municipality

	Net inflow 2017 (R mil.)		*Net Inflow 2018 (R mil.)*	
	Low	*High*	*Low*	*High*
Total outflow	−2383	−2949	−2199	−2723
Total inflow	4157	5449	3877	5169
Net inflow	1774	2500	1678	2445
As % of GDP	38%	53%		

Table 6.3 Net financial inflows: National

	Net inflow 2017 (R mil.)		*Net Inflow 2018 (R mil.)*	
	Low	*High*	*Low*	*High*
Total inflow	2383	2949	2199	2723
Total outflow	−1464	−1919	−1365	−1820
Net inflow	919	1030	834	903

6.3.3 *National economic analysis*

For the economic analysis, total outflows are higher since they take into consideration costs not addressed in the financial analyses, and total inflows are lower since financial flows from the insurance industry are not an economic benefit. Net inflows are negative and range between R2.6 and R3.1 billion (see Table 6.4). The Knysna fires had a negative impact on the national GDP of between 0.22% and 0.27% in 2017.

6.3.4 *Benefits of ecological restoration*

The benefits of the removal of invasive alien plants (in other words, the exiting of commercial forestry plantations in the region of the Knysna estuary) are the values fires averted (protection of residential and commercial property, and protection from loss of value of agricultural production) in 2017 and 2018, and also the annual benefit of ecosystem services restored. The costs are the costs of restoration (removal of invasive alien species, in other words the commercial pine plantations, the removal of the burnt and dead trees) and annual loss of timber values. Table 6.5 summarizes the results of the cost-benefit analysis.

Table 6.4 Net economic inflows: National

	Net inflow 2017 (R mil.)		*Net Inflow 2018 (R mil.)*	
	Low	*High*	*Low*	*High*
Total outflow	−2670	−3236	−2214	−2738
Total inflow	99	99	0	0
Net inflow	−2571	−3138	−2214	−2738
As % of GDP	−0.22%	−0.27%		

Table 6.5 Value of ecological restoration in the Knysna estuary

	Low	*High*
Net present value (R million, 2017 values, 6% discount rate, 20 years)	R 3,704.62	R 4,705.39
Benefit-cost ratio	6.21	7.62
Annualized value (US$/hectare/year)	$ 1,281.69	$ 1,627.93

6.4 Discussion

It matters at what level the impact of a fire is assessed, whether at a local or national level, and whether the analysis is purely financial or from an economic perspective. At a local level, the Knysna fires are projected to increase GDP by between 40% and 50% over two years. This is indicative of the economic benefits of restoration (i.e. clearing and removing invasive alien plants, which is what the fire achieved). At the same time, on a national level, the fire had an economic cost of between R2.6 and R3.1 billion after taking into consideration ecosystem losses and human health impacts. These costs decrease ecosystem values of the Knysna estuary (property, tourism values, and recreational values). The total value of restoration would therefore incorporate the local level benefits of restoration, plus the additional cost at the national level. To avoid double counting, the total cost comprises only costs in terms of human health as well as loss of ecosystem services values (as per the Swartkops case study, Chapter 2).

It is important to note that while certain sectors benefit from a fire, such as the construction industry and related sectors such as transportation and materials suppliers, other industries such as agriculture and the timber industries suffer. Even at a local level, it raises serious questions around the location of plantations and agricultural areas, and the importance of not producing these in fire prone areas. Also, the importance of clearing invasive alien plants has already been highlighted and recognized as a priority, if fires of this nature and intensity are to be avoided in the future.

References

Admin 2017. *Knysna municipality is in rebuild phase.* Knysna: Knysna Municipality. Available at http://www.knysna.gov.za/news/knysna-municipality-is-in-rebuild-phase (accessed on 12 October 2017).

BTE 2001. *Economic Costs of Natural Disasters in Australia*, Report No. 103. Canberra: Bureau of Transport Economics.

Citizen, The 2017. Fire death toll rises while blazes contained in Knysna, Sedgefield, Bitou and George. 11 June. Available at https://citizen.co.za/news/south-africa/1536261/live-knysna-fire/ (accessed on 26 January 2018).

De Villiers, J. 2017. "Knysna fire led to largest deployment of firefighting resources in SA history - authorities." News24, 20 June. Available at http://www.news24.com/southafrica/news/knysna-fire-led-to-largest-deployment-of-firefighting-resources-in-sa-history-authorities-20170620 (accessed on 11 October 2017).

Garden Route Rebuild 2017. Statistics of the Knysna Inferno. Available at https://www.gardenrouterebuild.co.za/2017/10/09/statistics-of-the-knysna-inferno/ (accessed on 17 October 2017).

Geldenhuys, C. 2017. "Die begin van'n brand [The beginning of a fire]." Rapport, 25 Junie.

George Herald 2017. The Great Fire of 1869. 17 August. Available at https://www.georgeherald.com/News/Article/General/the-great-fire-of-1869-20170817 (accessed on 12 December 2017).

Groenewald, Y. 2017. "Knysna fires, Cape storm devastating to insurance sector - Santam." Fin24, August 31. Available at http://www.fin24.com/Companies/Financial-Services/knysna-fires-cape-storm-devastating-to-insurance-sector-santam-20170831 (accessed on 12 October 2017).

Head, T. 2017. "Knysna Fires update: Official death toll in the region rises to six." The South African, 10 June. Available at https://www.thesouthafrican.com/knysna-fires-update-official-death-toll-in-the-region-rises-to-six/ (accessed on 26 January 2018).

Higgins, S.I., Turpie, J.K., Costanza, R., Cowling, R.M., Le Maitre, D.C., Marais, C. and Midgley, G.F. 1997. "An ecological economic simulation model of mountain fynbos ecosystems: dynamics, valuation and management." *Ecological Economics* 22(2):155–169.

Mitchley, A. 2017. "Donations flood in for #Knysna fire relief." News24, 9 June. Available at http://www.news24.com/SouthAfrica/News/donations-flood-in-for-knysna-fire-relief-20170609 (accessed on 12 October 2017).

Mortlock, M. 2017a. "Fire-hit Knysna gets rid of alien vegetation to prevent another disaster." Eyewitness News, 6 December. Available at http://ewn.co.za/2017/12/06/fire-hit-knysna-gets-rid-of-alien-vegetation-to-prevent-another-disaster (accessed on 12 December 2017).

Mortlock, M. 2017b. "Knysna fire victims won't have to pay rent until 2018." Eyewitness News, 2 September. Available at http://ewn.co.za/2017/09/02/knysna-fire-victims-won-t-have-to-pay-rent-until-2018 (accessed on 12 December 2017).

Payscale 2017. Entry-Level Fire Fighter Salary (South Africa) URL: http://www.payscale.com/research/ZA/Job=Fire_Fighter/Salary/01e2fd71/Entry-Level (Accessed on: 26 January 2018)

Smith, C. 2017. "Cost of Cape storm, Knysna fires likely billions more than R4bn – chamber." Fin24, 12 June. Available at http://www.fin24.com/Economy/cost-of-cape-storm-knysna-fires-likely-billions-more-than-r4bn-chamber-20170612 (accessed on 12 October 2017).

Stehle, T. 2017. "Knysna's Great Fire of 2017." SA Forestry Journal. Available at: http://saforestryonline.co.za/articles/knysnas-great-fire-of-2017/ (accessed on 11 October 2017).

Stephenson, C. and Handmer, J. 2012. "Black Saturday Costs." *Fire Australia Summer*(2011/12): 24–26.

Turpie, J.K. 2003. "The existence value of biodiversity in South Africa: How interest, experience, knowledge, income and perceived level of threat influence local willingness to pay." *Ecological Economics* 46(2): 199–216.

Turpie, J.K., Heydenrych, B.J. and Lamberth, S.J. 2003. "Economic value of terrestrial and marine biodiversity in the Cape Floristic Region: Implications for defining effective and socially optimal conservation strategies." *Biological Conservation* 112(1): 233–251.

Turpie, J.K., Forsythe, K.J., Knowles, A., Blignaut, J. and Letley, G. 2017. "Mapping and valuation of South Africa's ecosystem services: A local perspective." *Ecosystem Services*, 27, pp.179–192.

Viljoen, N. 2017. *Post-fire impacts of the Knysna-Plettenberg Bay Fires*. Knysna: Eden District Municipality, Department of Community Services.

Van der Merwe, M. 2017. "Knysna Blaze: Hunt is on for the fire starter." Daily Maverick, 15 August. Available at https://www.dailymaverick.co.za/article/2017-08-15-knysna-blaze-hunt-is-on-for-the-fire-starter/#.Wd42YCYaKcw (accessed on 26 January 2018).

Appendix A. Data used for estimating the financial and economic flows

Table A.1 Above- and below-the-line impact categories

Flow category	*Value 2017 (R mil.)*		*Note / Source*
	Low	*high*	
Above-the-line impacts			
Private residences	4000	5000	1
Infrastructure (incl. roads, water, public buildings)	136	185	2
Other public sector reconstruction	147	147	3
Rates exemption property owners	16.5	16.5	3a
Total urban	4283	5331	
Agriculture (emergency measures, excl. feed)	30	30	4
Cost of fodder destroyed	10	10	5
Agriculture (loss of food production)	7	7	6
Timber (plantation) loss of production	50	50	7
Emergency response operations (salaries only)	1	43	8

(*Continued*)

Table A.1 (Continued)

Flow category	*Value 2017 (R mil.)*		*Note / Source*
	Low	*high*	
Fatalities	***254***	***254***	***9***
Injuries			
Serious	***18***	***18***	***10***
Minor (firefighters only)	***1***	***1***	***11***
Environment – loss of ecosystem services			
Fynbos	***4***	***4***	***12***
Natural forest	***7***	***7***	***13***
Azonal	***4***	***4***	***14***
Environment – costs of restoration	135	135	15
Below-the-line impacts			
Western Cape Provincial Government	75	75	16
Knysna Fire Disaster Relief Fund	1.3	1.3	17
Insurance claims (estimated)	2730	3640	18
Other donations (not exhaustive)	22	22	19
National government contributions	??	??	

Notes: Economic impacts italicized and also in bold. The rest are financial impacts.

1 Value: De Villiers (2017)

2 Value: De Villiers (2017) [low]; Van der Merwe (2017) [high]

3 Value: Van der Merwe (2017)

3a Value: Mortlock (2017b)

4 Value: Van der Merwe (2017)

5 Value: Garden Route Rebuild (2017)

6 Unit cost: Turpie et al. (2017); Area: Viljoen (2017)

7 Value: own calculation

8 Salaries: Payscale 2017. Number: De Villiers (2017); Fire blazed for 6 days (7–12 June = 144 hrs), roughly 1 000 firefighters deployed (Stehle 2017).

9 Unit cost: Stephenson and Handmer (2012); Number: Viljoen (2017). Use lowest pay grade since some will be volunteer firefighters whereas others receive higher salary. Usually work shifts so not all deployed over full period. However, will stay longer than duration of fire for mop up work.

10 Injuries estimate very crude. Unit costs: BTE (2001). No. affected: Head (2017). There were six dead and five critically injured at the time. The official death toll was 7, so we assume one of the critically injured died of their injuries.

11 Unit costs: BTE (2001). No. affected: Citizen (2017)

12 Unit cost: Turpie (2003) based on Higgins et al. (1997) [upper bound] [water production, direct use values, genetic storage]. Area: Viljoen (2017).

13 Unit cost: Turpie et al. (2003); Turpie (2003); Turpie et al. (2017) [Direct + existence + genetic + carbon storage]. Area: Viljoen (2017)

14 Unit cost: Turpie et al. (2017) [tourism+sediment retention+flow regulation+water quality amelioration]. Area: Viljoen (2017)

15 Value: Van der Merwe (2017)

16 Admin (2017)

17 Admin (2017)

18 Insurance costs for Knysna fires and Cape storms between R3 and R4 billion (Smith 2017). Santam paid R800 million, of which R72 million related to Cape property damage (Groenewald 2017). Use ratio to estimate share of insurance claims for Knysna fire. Not all of these claims are paid. May be too early to assess this accurately.

19 Mitchley (2017). As at 9 June 2017. Donations to variety of agencies e.g. Gift of the Givers.

7 Discussion and summary of findings

7.1 Introduction

Until now, the focus has been on evaluating specific economic components in certain regions. In Chapter 2, the costs of the degradation of Swartkops estuary were estimated in terms of the impact on recreation and human health. In Chapter 3, the economic benefits of restoration were estimated in terms of improved water quality on four ecosystem services (property, nursery, subsistence and recreation). Chapters 4, 5 and 6 considered the economic benefits associated with the removal of invasive alien plant species in Swartkops, Great Brak and Knysna estuaries, respectively. In this chapter, the results of the study are contextualized with references to other studies in the literature.

7.2 Literature review

The economic contribution of estuaries in South Africa is fairly well studied, although most studies have understandably focused on the recreational value of estuaries. These values are sizable, ranging from R609/hectare to R3.4 million/hectare (Table 7.1). It should be noted that those estuaries with high values are those that are regularly frequented by tourists. Many estuaries in South Africa are less frequently visited and would consequently have a lower recreational value. So, when averaging across all estuaries in South Africa, one would expect the mean recreational value to be much lower. The methods used to estimate recreational values also influence the value, as stated (CVM) and revealed (TCM) preference methods measure the area under the demand curve, whereas methods based on actual expenditure typically result in much lower estimates (see Chapter 2 for an elaboration of this).

A number of studies have also estimated the value of different ecosystem components of all estuaries in South Africa. McGrath et al. (1997) estimated that the estuarine fishing industry and the

DOI: 10.4324/9781032651675-7

Table 7.1 Review of studies that estimated the recreational value of estuaries (in 2019 Rands/hectare)

Study	*Site*	*Method*	*Value*
Cooper et al. 2003	Berg Estuary	TCM	913,920
	Breede Estuary	TCM	2,003,958
	Umhlatuze Estuary	TCM	5,359
	Keiskamma Estuary	TCM	240,326
Turpie et al. 2005	Knysna Estuary	TCM	124,483
		TCM	3,368,035
Hosking and du Preez 2004	Keurbooms	CVM	888,045
Hosking and du Preez 2004	Keurbooms Estuary	CVM	3,404,172
Mann et al. 2002	St Lucia Estuary	Spend	609
Hosking 2011	Bushmans Estuary	CVM	31,397
Pradervand 1998	Kromme + Great Fish	Spend[1]	91,393

Source: adapted from Nahman 2006.
Notes: CVM = contingent valuation method; TCM = travel cost method; Spend = actual expenditure

[1]running costs only

estuarine-dependent (as a nursery) marine inshore fishery were worth in the region of R1 billion per annum (1997 price levels). In 2019 prices, the contribution of estuaries to fishing and marine dependent nurseries is estimated to be in the region of R3.34 billion per annum. Lamberth and Turpie (2003) estimated the value of estuaries for fishing and inshore marine. They estimated that the value of all estuaries was R1.251 billion (in 2002 Rands), or R3.058 billion in 2019 Rands. This equates to R29,546/hectare for all estuaries.

At the same time, the values of estuaries in urban areas are much higher than those in rural areas. In a study for Durban Bay estuary reported on in Mander (2001), the value of "free services" such as water purification and incomes generated from estuary activities amounted to R150 million per year.

These values, however, do not include other ecosystem values, such as nursery, recreation & tourism values, and existence values. More recently, Turpie and Clark (2007) estimated these value. They determined that the total value for all estuaries in South Africa was R3,228 million per year in 2007 values (or R5,488 million per year in 2019 Rands), but although property values of estuaries were assessed, the total value of estuaries reported by them did not include property values. This is in one sense reasonable, since development may actually detract from the value of the estuary.

Property values in particular are likely to be substantial. Turpie et al. (2017) estimated that the property value of coastal areas is R79 billion per year (2015 prices) or R96 billion per year in 2019 Rands. Combining these values, the total value of estuaries in South Africa is therefore estimated to be over R100 billion per year in 2019 prices (property, recreational, nursery, existence, and subsistence values). This equates to R966,184 per hectare per year for all estuaries (2019 values). Comparing this to an estimate by Turpie et al. (2017) of the total economic value of ecosystem services in South Africa (R275 billion per year in 2015 prices, or R335 billion per year in 2019 prices), indicates that estuaries potentially contribute 30% of ecosystem values in South Africa, a substantial contribution. However, the recreational values reported in Table 6.1 and the discussion above suggest that the value of some estuaries could be substantially higher than this.

No studies have estimated the value of restoration in South Africa. Opaluch et al. 1999 estimated the average per hectare dollar values (in 2005 prices) were $14,400 for unpolluted shellfish grounds, $15,360 for saltmarsh, and $19,2000 for eelgrass beds in the Peconic Estuary System (PES) in New York. Studies like this indicate the potential values associated with restored estuarine habitats. Johnston et al. (2002) developed a simulation model for this system.

Cordier et al. (2017) developed a simulation model to assess the economic impacts of restoration on an estuary in France. They showed that the costs of estuary restoration negatively impact Gross Domestic Product (GDP). However, they did not quantify the value of ecosystem service improvements, in contrast with the studies reported previously. Their focus was purely on the real economy. Their study is nonetheless important as it represents one of the first papers to model the improvements in estuaries as a result of ecological restoration using a system dynamics model.

7.3 Method

The study utilized a variety of methods. In Chapter 2 (Swartkops) and Chapter 6 (Knysna estuary), the change in productivity method was used to assess the impact of declining water quality on fish productivity, and the cost of illness and human capital approach was used to estimate the impacts of declining water quality on human health using the actual expenditure method.

In Chapters 3, 4 and 5, a cost-benefit analysis was considered comparing the benefits in terms of additional ecosystem values associated with ecological restoration of estuaries. These were compared with the economic costs thereof. Because we were interested in a sociological systems approach for analysis of benefits and costs (Adams et al. 2020), three system dynamics models were developed that considered:

i improvements in water quality through refinements to wastewater treatment works at Swartkops estuary
ii the removal of water hyacinth in Swartkops estuary, and associated improvement in seagrass
iii the removal of smooth cordgrass (*Spartina alterniflora*) and the associated improvement in native marsh species

System dynamics models are well established in the literature for modelling complex systems characterized by non-linearity and feedbacks (Sterman 2000).

7.4 Summary of results

Turpie (2007) estimated that a partial protection of 80% of estuaries would increase the value over the existing value of all estuaries by R346 million. Here, it is evident that ecological restoration could have a large impact on economic values. The improvement of estuarine water quality alone contributes almost R90 million per year to the Swartkops estuary (in 2019 terms, see Table 6.2).

Costs of illness and other productivity effects could reduce the estuary value by 4–6%, but note that this value is based on the expenditure methodology, which was shown to generate values that are much lower than the contingent valuation method, which estimates the total area under the demand curve. Furthermore, the travel cost method was found to report values that are 20–30% higher than those produced by the contingent valuation method (see Winpenny 1995).

Furthermore, the Swartkops estuary is a highly degraded system; therefore, the costs reported here are lower bound, and they do not include a number of other impacts. Le Roux et al. (2005), for example, estimated a total cost of 20% of value for Kongweni estuary (see Chapter 1). In spite of the limitations on the cost component of the study, it was found that restoration could improve the ecosystem values associated with the Swartkops estuary by 16%. This highlights the importance of restoration in these ecosystems. Pouso et al. (2019) estimated that ecosystem restoration could improve the current values by 7.5–11.5%. These results show the important role that restoration can play in the estuarine environment.

Chapter 6 shows that ecological restoration following a fire can produce substantially greater improvements in value. In the Knysna estuary, following the Great Fire of 2017, restoration could potentially improve the value of ecosystem services (and mitigate the impact of fires on property values and agricultural productivity) by increasing the value of the estuary by 7% over value, after taking into account degradation (Table 7.2). At the other end of the spectrum, the Great Brak estuary shows a moderate

Table 7.2 Summary of values, costs of degradation and benefits of restoration, for the different estuaries in this study (Rand million, annual 2019 values)

	Swartkops		*Great Brak*		*Knysna*	
Source	This study		This study		This study	
Value	470	Turpie and Clark 2007, subsistence, property, tourism, nursery	833	Turpie and Clark 2007, subsistence, property, tourism	5,059	Turpie and Clark 2007, subsistence, property, tourism, nursery
Cost of degradation	20.2	Lower bound (Chapter 2), cost of illness and change in productivity	n/a		313	Lower bound (Chapter 6), human health impacts and loss of ecosystem services
(% of value)	4%				6%	
Net value	449.8				4,746	
Restoration:						
A. Improved water quality and reduced flow from WWTW (50%)	89.40	Lower bound (Chapter 3), change in TEV of estuarine habitat	n/a			
B. Removal of invasive plant species	0.44	Chapter 4	−0.03	Chapter 5 (value of seagrass recovery) baseline	352.09	Lower bound, Chapter 6
Total value	559.85		832.97		5,411.05	
Improvement over initial values	16%		0.00%		7%	
Value of restoration (2020 US$/ha/year)	6,058		−1,334		1,223	

Note: change in fishing values as a result of changes in water quality (R13.6 million) removed from the total value to avoid double counting.

(negligible) decline in value (0.00%), even though only the removal of Spartina spp is considered. For the Kongweni estuary in KwaZulu-Natal, it is estimated from data in Chapter 1 that the improvement in estuary values (recreational benefits) as a result of an improvement in water quality improvements results in an increase over initial estuary values (after degradation) of 22% (see Le Roux et al. 2005).

We can, therefore, conclude that the benefits of ecological restoration of estuaries are much higher than were previously supposed. Estuarine environments provide important ecosystem services and are often degraded; therefore, restoration has huge potential to increase the economic values associated with these systems.

7.5 Conclusions

Estuaries made a sizable contribution to the economic values of ecosystems in South Africa, but there are even greater potential economic benefits from restoring estuaries. Restoring estuaries to pristine levels could potentially increase ecosystem values by between 0.00% and 22%, based on this study and a review of other studies in the local and international literature.

Although the results are still very preliminary, the economic benefits from estuary restoration are likely to be substantial, much higher than ecological restoration in catchments and other terrestrial areas.

References

Adams, J.B., Whitfield, A.K. and Van Niekerk, L. 2020. "A socio-ecological systems approach towards future research for the restoration, conservation and management of southern African estuaries." *African Journal of Aquatic Science* 45(1–2): 231–241.

Cooper, J., Jayiya, T., Van Niekerk, L., De Wit, M., Leaner, J. and Moshe, D. 2003 *An assessment of the economic values of different uses of estuaries in South Africa*. Stellenbosch: CSIR Environmentek.

Cordier, M., Uehara, T., Weihe, J. and Hamaide, B. 2017. "an input-output economic model integrated within a system dynamics ecological model: Feedback loop methodology applied to fish nursery restoration." *Ecological Economics* 140: 46–57.

Hosking, S.G. and Du Preez, M. 2004. "A recreational valuation of the freshwater inflows into the Keurbooms estuary by means of a contingent valuation study." *South African Journal of Economic and Management Sciences* 7(2): 280–298

Hosking, S. 2011. "The recreational value of river inflows into South African estuaries." *Water SA* 37(5): 711–718

Johnston, R.J., Grigalunas, T.A., Opaluch, J.J., Mazzotta, M. and Diamantedes, J. 2002. "Valuing estuarine resource services using economic and ecological models: The Peconic Estuary System study." *Coastal Management* 30: 47–65.

Lamberth, S.J. and Turpie, J.K. 2003. "The role of estuaries in South African fisheries: economic importance and management implications." *African Journal of Marine Science*, 25, pp. 131–157.

Le Roux, R., Nahman, A., Pillay, S., Weerts, S. and Reyers, B. 2005. *The economic impacts associated with a change in the environmental quality of the Kongweni Eastuary at Margate, KZN*. CSIR, Report No. ENV-SC 2005, 51.

Mander, M. 2001. "The value of estuaries." In Breen, C.M. and MeKenzie, M. (Eds). *Managing estuaries in South Africa: An introduction*. Pietermaritzburg: Institute of Natural Resources.

Mann, B.Q., James, N.C. and Beckley, L.E. 2002. "An assessment of the recreational fishery in the St Lucia estuarine system, KwaZulu-Natal, South Africa." *South African Journal of Marine Science* 24: 263–279

McGrath, M.D., Horner, C.C.M., Brouwer, S.L., Lamberth, S.J., Sauer, W.H.H. and Erasmus, C. 1997. "An economic valuation of the South African linefishery." *South African Journal of Marine Science* 18: 203–211.

Nahman, A. 2006. *Valuing water quality changes in the Kongweni estuary, South Africa, using contingent behaviour data*. M.Sc. dissertation, University of Manchester, UK.

Opaluch, J.J., Grigalunas, T.A., Diamantedes, J., Mazzotta, M. and Johnston, R.J. 1999. *Recreational and resource economic values for the peconic estuary system*. Economic Analysis Inc.,. February, 1999. Peacedale, Rhode Island

Pouso, S., Ferrini, S., Turner, R.K., Borja, A. and Uyarra, C.M. 2019. "Monetary valuation of recreational fishing in a restored estuary and implications forfuture management measures." *ICES Journal of Marine Science* 9: fsz091. doi: 10.1093/icesjms/fsz091

Pradervand, P. 1998. *An assessment of recreational angling in Eastern Cape Estuaries*. MSc thesis, University of Port Elizabeth. Port Elizabeth.

Sterman, J. 2000. *Business dynamics: systems thinking and modeling for a complex world*. New York: McGraw Hill.

Turpie, J. and Clark, B. 2007. "Development of a conservation plan for temperate South African estuaries on the basis of biodiversity importance, ecosystem health and economic costs and benefits. C.A.P.E. Regional Estuarine Management Programme." Final Report; August 2007. 125pp.

Turpie, J. 2007. *Guideline 9: Maximising the economic value of estuaries. Cape Action for People and the Environment estuaries programme*. Cape Town: Anchor Environmental consultants.

Turpie, J.K., Forsythe, K.J., Knowles, A., Blignaut, J. and Letley, G. 2017. "Mapping and valuation of South Africa's ecosystem services: A local perspective." *Ecosystem Services* 27: 179–192.

Turpie, J.K., Clark, B., Napier, V., Savy, C. and Joubert, A. 2005. *The economic value of the Knysna estuary, South Africa*. Report submitted to Marine and Coastal Management, Department of Environmental Affairs and Tourism, Cape Town.

Winpenny, J.T. 1995. *The economic appraisal of environmental projects and policies – A practical guide*. Paris, France: OECD & ODI.

Index

Printed in the United States
by Baker & Taylor Publisher Services